OBJECTIVE QUESTIONS ON "VEGETABLE WORLD"

A Right Approach towards Success in JRF/SRF/ARS/ NET/ SAU Entrance /PCS Examinations

Includes Previous Years' Questions of above Examinations (Memory Based)

J.P. SRIVASTAVA
S. LAL
GAURAV MISHRA

MD PUBLICATIONS PVT LTD
New Delhi

Published by

MD Publications Pvt Ltd

"MD House", 11, Darya Ganj, New Delhi-110002

Tel. : +91-11-41562846, 41563325

E-mail : contact@mdppl.com

Website : www.mdppl.com

ISBN : 978-81-7533-129-7

PRICE : Rs. 495.00

Printed and Published by **MD Publications Pvt Ltd** at Times Press, New Delhi.

PREFACE

Vegetables are very important crops from nutritional point of view and source of income to the farmers. These crops are commercial and this is the reason that these are grown near the cities and towns. Simple grading and processing add to the value of these crops. On account of these reasons these crops are taught to M.Sc. (Ag.) in State Agricultural Universities and ICAR Institutes. Also, there is a section on vegetable crops in Agricultural Research Service Examination. The questions on vegetables are also frequently asked in ARS, JRF, SRF, NET, IAS and PCS examinations. A vast knowledge has been gathered by the students on vegetables, but they are not up-to-date in quickly replying the objective type of questions. Therefore, efforts were made to set objective type questions along with their answers in different groups of vegetables and the same have been arranged in this book. Several faculty members of vegetable science and the authors are involved in this compilation. We hope that this book will be useful to the candidates testing their luck in ARS, JRF, SRF, NET, IAS and PCS examinations.

The readers are welcomed to pointout mistakes, errors and omissions, if any and give their valuable suggestions for improving the quality of this book.

(J. P. Srivastava)

PREFACE

Prof. (Dr.) V. K. Suri
B.Sc. Agri. Hons., M.Sc. Soils (PAU). Ph.D. (IIT) FISAB
Vice-Chancellor

C. S. Azad University of Agriculture & Technology
Kanpur-208 002, U.P., India

FOREWORD

Vegetables are the most important crops from nutrition point of view as they are rich source of vitamins and minerals, and are also of medicinal importance. This is the reason that WHO has recommended vegetable intake of atleast 300 g/caput/day in the diet.

In vegetable production, India stands second in the world, first being the China.

The vegetables give lucrative economic to the farmers. In order to ensure regular supply of vegetables, special storage and marketing/transportation facilities have been created in the country. Not only the Polyhouses have been designed, the progressive farmers have ventured into off-season vegetable production through such facilities. Rural state agricultural universities and ICAR Institutes have started research and teaching programmes at M.Sc. and Ph.D. Ievels in vegetables. Private companies, including multinational and NGO have also embanked upon varietal development, mostly hybrids and seed production.

In order to promote research and teaching on vegetables scientists at S1 (Assistant Professor) level are recruited by Agricultural Scientists Recruitment Board at national level through competitive examinations. Often, questions on vegetables are asked in ARS, JRF, SRF, NET, IAS, PCS, UGC, etc., examinations. Therefore, there was an urgent need to have a book on objective type questions on vegetables alongwith their keys. Efforts made by Dr. J.P. Srivastava, Head, Vegetable Science, Dr. Shankar Lal, former Vegetable Breeder and Ex-Director, Indian Institute of Pulses Research and Sri Gaurav Mishra, Research Scholar, C. S. Azad University of Agriculture and Technology, Kanpur are praiseworthy and I hope that this book would be useful to the students entering competitive examinations seeking admission into M.Sc. and Ph.D. Programmes in vegetables as well as graduate students appearing in ARS, PCS, IAS examinations. I once again congratulate the authors of this book for their efforts and I hope that they will continue such efforts in future as well as

(V. K. Suri)
Vice-Chancellor

ABOUT THE AUTHORS

Dr. J.P. Srivastava

Dr. Srivastava has obtained his B. Sc. (Ag.) and M. Sc. (Ag.) degrees from Govt. Agriculture College, Kanpur (Now C. S. Azad University of Agriculture & Technology) and Ph.D. degree from Kanpur University, Kanpur in 1972. After serving in various capacities at Vegetable Research Station, Kalyanpur-Kanpur, he is now heading the Department of Vegetable Science of C. S. Azad University of Agriculture & Technology, Kanpur.

Dr. Srivastava is a renowned vegetable breeder and has developed more than 50 varieties / hybrids of various vegetable crops, some of them like Azad P-1, Azad P-3 in table pea, Kalyanpur Baramase in bitter gourd, Azad T-5 and Angoorlata in tomato, Azad hybrid in brinjal, Azad Mirch-1 in chilli etc., are nationally popular. He has also given much significant contribution in research, teaching and extension related to vegetable crops under National Seed Projects of World Bank. He has made remarkable contribution due to which both the farmers and the University have been benefited monetarily. He is credited by 65 research papers, 110 abstracts of symposia/seminars and more than 100 popular articles on Vegetable Production Technology. Dr. Srivastava has been Principal Investigator of several research projects funded by national and international organisations. He has also guided research to several M. Sc. (Ag.) and Ph.D. students in the field of vegetable science.

S. Lal

Dr. Shankar Lal is a renowned breeder, teacher and research manager. He retired as Director of Indian Institute of Pulses Research, Kanpur. After obtaining M. Sc. (Ag.) and Ph.D. degrees from Govt. Agricultural College (Now C. S. Azad University of Agriculture and Technology, Kanpur), he served the same college as teacher in agricultural Botany and Vegetable Breeder. After serving Govt. of U.P. as officer-in-charge, State Soil Conservation and Training Centre, Rehmankhera, Lucknow as Economic Botanist (Legumes), he joined Indian Agriculture Research Institute, New Delhi and from where he joined Haryana Agriculture University, Hisar as Professor of Plant Breeding (Pulses). From this position, he turned as Research Manager at N.D. University of Agriculture and Technology, Faizabad, Asstt. Director General (Food & Fodder crops), ICAR and Director, Indian Institute of Pulses Research, Kanpur.

Dr. Lal has published more than 75 research papers in journals of national & international repute, more than 200 popular articles and about 10 bulletins and books. He has contributed in crop improvement by developing 50 varieties of pulses and 6 varieties of vegetables. He has more than 25 years of teaching experience in the field of genetics, plant breeding and seed production. He has extensively traveled both within and outside of the country (Australia, Syria, Thailand, Philippines, Nepal and Bangladesh).

Dr. Gaurav Mishra

Dr. Gaurav Mishra is a young scientist in the field of vegetable science. He has obtained B.Sc. (Ag.) and M.Sc. (Ag.) degrees from CSJM University and CSAUAT, Kanpur respectively. He has been selected to be awarded as "Chancellor Gold Medal" from CSAUAT, Kanpur. Recently, he has obtained Ph.D. degree in vegetable science. Dr. Mishra has published 13 research papers, 25 abstracts and more than 25 popular articles. Currently, he is serving in State Bank of India as "Technical Officer (Farm sector)" for promoting agriculture business through monitoring the projects to be financed by the Bank. He is also providing technical support to the operating Branches besides finalizing Agricultural projects.

CONTENTS

CHAPTER 1

GENERAL INFORMATION ON VEGETABLES

Vegetables play a vital role in the health and nutritional security of human beings in addition to improving the economy of the farming community of the country.

According to FAO, about 25-30 million people die from hunger and exhaustion every year globally. In India adverse effects of chronic malnutrition are of greater magnitude in low-income group. Obviously, the socio-economic level, especially purchasing power of the family, dictates the quantity and quality of food for the family. Farmer families in India, who grow the food for the country, are more poorly fed than many slum dwellers. Nearly, half of all pre-school children in the developing countries are deprived of normal nourishment which causes various physiological disorders and at the same time physical and mental retardation. In India, majority of the pre-school children are inadequately nourished. According to World Health Organization, xerophthalmia (dryness of the eyes due to deficiency of vitamin A) afflicts nearly 13-15 million infants and pre-school children every year. At least, half million of them become partially or totally blind each year and approximately two-thirds die within a few months of becoming blind. In the developing countries, where 52% of the world's arable land is concentrated and 74% of the world population lives, only 40% of the world's food stuff is produced. Under the circumstances of hunger and under nourishment, particularly in the developing countries, vegetable crops can play a very crucial role in alleviating this malady, because these crops can give very high tonnage of protective foods. According to FAO, most of the developing countries fall under the category of countries with insufficient per capita vegetable consumption, under 25 kg/annum, where dieticians recommended that a person should consume 100-120 kg vegetables per year (300-350 g/day).

Now let us examine the diet composition in our country. Cereal-based diet is the main feature of Indian food habit. We achieved a great degree of stability in food grain production. Average per capita availability of food grain is estimated to be around 400 g per day for the last 10 years. Still about 19 crores of Indians, according to WHO, are half-fed, because the food grains which are mostly cereals constitute only a part of the diet. The extensive studies carried out by the Indian Council of Medical Research, New Delhi and the

National Institute of Nutrition, Hyderabad revealed that meager intake of low cost protective foods, particularly vegetables are largely responsible for malnutrition among the majority of our population. Chief deficiencies in our diet are calories, proteins, vitamin A and riboflavin. In fact, the present per capita supply of both calories and proteins in India is probably the lowest in the world. The key to improving the nutrition of the rural people in mostly vegetarian India is to provide low cost protective foods, like vegetables.

A balanced diet i.e. well-planned diet which provides enough food and ensures at least a minimum supply of carbohydrates, proteins, fats, vitamins and minerals to keep the physiological processes of the body in good tune. According to the dieticians, the daily balanced diet for an adult in our condition should constitute 125 g leafy vegetables, 75 g other vegetables and 100 g root and tuber vegetables along with 475 g cereals, 80 g pulses, 92 g fruits, 200 g milk, 40 g oil and 40 g sugar or jaggery. In non-vegetarian diet, 40 g meat or fish and 30 g egg are added and the quantity of pulses and milk is reduced to 65 g and 100 g, respectively. Do, importance of vegetables in both the vegetarian and non-vegetarian diets remain the same. Many vegetables, especially tuberous ones, can quite successfully constitute the bulk of the seed stuff. In many developing countries, national agricultural development programmes are focused on the need to increase the production of vegetables in order to increase their proportion in the daily diet. Now let us see how much vegetables are available in our daily diet at present. Our annual requirement of vegetables is likely to go up to 150 million tonnes by 2010 AD for an estimated population of 126 crores. At present, 109 million tonnes of vegetables (2005—2006) are being produced annually from an area of 7.1 million hectares (excluding potato). Even this low level of average supply does not fully reflect the consumption pattern of the rural households and those below the poverty line where per capita vegetable consumption is very low, even lower than 100 g per day. This situation can be understood from a national level sample survey which reveals that 71% Indians spend 82% of their income for purchasing only food grains and remaining 18% for the other food stuffs and essential commodities. Low production of vegetables mainly due to low productivity is the prime reason for such meager availability in our diet. Intake of vegetables is further reduced due to colossal post-harvest losses which are estimated to have been about 35-40 per cent. Moreover, hardly 3.0% of our total cropped *area* is under vegetable cultivation. It is now well conceived that by simply adding greens and other vegetables to the available 400 g food grains, the diet of the average Indians can substantially be upgraded. This realization had been amply reflected in the Eighth five year plan where annual vegetable production targets were fixed at 75 million tonnes from 6.0 million hectares for short term (1990-95) and at 150 million tonnes from 8.0 million hectares for long term (2010 AD) basis.

In India, per capita land resources are limited (0.12 ha) which will further be decreased due to tremendous growth of population. Under such circumstances, it is very important to obtain maximum production per unit area within the shortest period of time. The vegetable crops best satisfy these requirements because their productivity is 5-10 times higher than that of any cereal crop. Gross yield of calories per hectare from vegetables is also comparable with that from cereals. For example, wheat yields, on an average, 8700 Kcal per hectare, whereas tomato yields 12,540,000 Kcal and cabbage 6,000,000 Kcal per hectare. In addition, increased vegetable cultivation creates additional employment in rural sector as it is a labour intensive farming practice. Vegetable crops have vast potential for export and processing industries.

India's export of fresh vegetables during 2003-04 registered a steep growth of 48.19 per cent over the previous year when the same reached a level of Rs 27.39 crores as against Rs. 11.82 crores. This was followed by onions (97.86%) and peas (11.36%), onions, as may be seen, countries to be the largest item of export. It alone accounted for 76.18 per cent in the total export of vegetables from the country during 2003-04. Other vegetables, next in importance, comprised cucumbers and gherkins, mixed vegetables (Potatoes and Mushrooms), different processed products of vegetables, dehydrated products and frozen vegetables.

Vegetable cultivation has a checkered history in India. Mention of different kinds of vegetables, like bottle gourd (*alabu*), (*kakinga*), onion (*palandu*), brinjal (*vartaku*), hyacinth bean and radish are available in ancient Sanskrit literature and in the epic like Ramayana. Vegetable cultivation took momentum during the Aryan invasion (1800 BC) and later developed by the Mughal rulers. It was flourished by the colonist traders from Portugal, France and Britain, who introduced some new vegetables, like cabbage, cauliflower, carrot, potato, knolkhnol, chilli, tomato, asparagus, etc., in India. At present, India is the second largest producer of vegetables in the world after China.

1.1 Diversity in vegetable cultivation

India grows a largest number of vegetable crops due to diversity in agroclimatic conditions. Nearly 60 kinds of fruit, leafy, root, bulb, starchy tuberous and other types of vegetables are cultivated in the country. The status of vegetable cultivation in India is unique, consisting of diverse kinds, such as, kitchen gardens of urban area, market gardens near big as well as metropolitan cities and truck farming of special crops in larger quantities for distant markets. Some of such truck farms are situated in Ranchi, Solan, Poona-Nasik belt, Punjab, West Bengal, Bangalore, etc. Produces of some specified areas are exported. These areas are Nasik, Ozar, Saikheda, Dindhori, Kolhar, Naraingaon and Sholapur in Maharashtra for okra; Pen, Alibaugh and Choul in Raigad district Dindhori and Niphad in Nasik district of Maharashtra for chillies; Dindhori, Niphad and lgatpuri taluka in Nasik district and Pen, Alibugh in Raigad district of Maharashtra for bitter gourd; Nasik, Pune and Satara districts, in Maharashtra, Periyar and Coimbatore in Tamil Nadu, Badaun in Uttar Pradesh and Patna, Bihar sharif in Bihar, Bhavanagar and Rajkot in Gujarat for big onion (discussed later under export of vegetables).

Vegetable processing industries require a particular cultivar or type of vegetables. Farming of vegetables for processing industries is prevalent in some pockets around processing centres located at Bangalore, Kolkata, Nagpur, Haldwani, etc. Some areas, like Hajipur (Bihar), Faizabad and Jaunpur (UP), Kullu and Saproon valley (HP) and Kashmir valley are devoted to the seed production of cauliflower, cabbage, carrot, beet, etc. Riverbed or 'Diara cultivation of cucurbitaceous vegetable crops is practiced along the rivers Jamuna, Ganga, Gomati, Saryu in Haryana, Uttar Pradesh and Bihar; Banas riverbed in Tonk district of Rajasthan; Narmada, Tawa, Tapti riverbed in Madhya Pradesh and Maharashtra; Sabarmati, Panam, Vartak and Orsuch riverbeds of Gujarat and Tungabhadra, Krishna and Pennar riverbeds of Andhra Pradesh. Growing of vegetable crops on a floating base prepared by a type of grass (*Typha* sp.) and organic matters over the Dal lake of Kashmir is really unique of its own kind.

1.2 Area, production and productivity of vegetable crops

In the recent past, tremendous progress has been made for increasing the vegetable production in the country, as a result India has emerged as second largest producer of vegetables being next to China. During the year 2000-2001, India has produced 93.922 million tonnes of vegetables from 6.249 million ha of land accounting a productivity of 15.0 t/ha and during the year 2003-04 it is estimated that the total vegetable production of country will touch the 1 billion tonnes. The vegetable production was very low, less than 20 million tonnes during 1947. The production of vegetables till 1961-65 was about 23.45 tonnes in 1967-71 and to 39.99 million tonnes in 1986. During 1991-92 the total production was 58.543 million tonnes from 5.593 million ha of land, productivity being 10.5 t/ha. Thus in last one decade country's vegetable production has increased more than 35 million tonnes and gross vegetable productivity of the country multiplied by one and half times. Presently India's share is 13.60% of total world production of vegetables and occupies second position next to China.

Due to advent of hybrid technology and general awareness of nutrition security among the people, vegetable production of the country is getting the right momentum. In last one decade production of vegetable pea has gone up more than 3.5 times while cabbage and tomato production has multiplied by 2.02 and 1.71 times, respectively (Table 1). Similarly, there has been substantial increase in the productivity of the vegetables. Due to availability of potential and high temperature tolerant varieties of cabbage, its productivity has gone up more than 1.46 times as compared to its productivity in 1990-91. In the country, Tamil Nadu has the credit to bag the highest productivity i.e. 27.5 tonnes/ha followed by Kerala (22.0 tonnes/ha) and Uttar Pradesh (19.5 tonnes/ha). Productivity in some of the states like Delhi, Rajasthan and all the states comprising north eastern hill region, except Tripura are with range of more than 10 tonnes per hectare. In these states there is urgent need to recast the vegetable production strategy so that vegetable production may be impetus. Further, for enhancing the vegetable production in the country our target should be to achieve 25 tonnes/ha productivity by 2006 which is presently 15 tonnes/ha.

Table 1.0: Change in growth of production and productivity in a decade (1991-92-2005-06)

Crop Parameter	1991-92		2005-06		Per cent increase in production	Per cent increase in productivity
	Production (000mt)	Productivity (t/ha)	Production (000 mt)	Productivity (t/ha)		
Brinjal	4612.2	15.3	7676.9	16.3	66.45	6.53
Cabbage	2771.2	15.6	5617.1	22.9	102.69	46.79
Cauliflower	2998.1	14.8	4694.5	18.3	56.58	23.64
Okra	1886.5	8.5	3344.6	9.6	77.29	12.94
Peas	851.6	7.1	3007.6	9.4	253.17	32.39
Tomato	4243.4	14.7	7277.1	18.4	71.49	25.17

1.3. All India Coordinated Research Project on Vegetable Crops

All India Coordinated Research Project on vegetable crops was started during the fourth five year plan in 1970-71, to provide a national grid for multi-location testing of the vegetable technologies developed by various research institutes and state agricultural universities. The headquarter of the project was at the Division of Vegetable Crops, Indian Agricultural Research Institute (IARI), New Delhi and the Project Coordinator (vegetable crops) joined the project in July 1971. During 1986, level of the project was elevated to the Project Directorate. Further, a significant development during 1992, headquarter of the Directorate was shifted from IARI, New Delhi to the Indian Institute of Vegetable Research (IIVR) at Varanasi. Presently the project has 26 centres located in different agro-climatic zones of the country. Besides these, about 30 centres involving central institutes, state and traditional universities and a few other public and private research organizations are participating in the cooperative research programme of the project on voluntary basis.

1.4 Organic farming of vegetable crops

Most of the vegetables are succulent and attract several insect pests. In order to control them, pesticides are used. These pesticides remain chemically active for a period and produce hazardous effects on the environment. Some of them do not degrade and enter in human body with the vegetables consumed by individuals. Some of these chemicals are carcinogenic, therefore, in advanced countries like USA and Europe, consumers demand for organically produced foods. Organic vegetable production is now a commercial venture, although on small scale, in several developed countries. In these countries, organically produced vegetables are available in the market at a premium price. Right from production to the shelf, there are strick checks and watch on the production technique, use of inputs and processing so that the extra money paid by consumer bring added satisfaction. Thus, the consumer demand for organically produced food is driven by environmental and food safety concern.

At the international level, a joint working group of WHO/FAO, Codex Alimentarius Commission develops guidelines to law makers, in participating countries. However, for organic products, 'basic standards' of international Federation of Organic Agriculture Movement (*FOAM*) serve the member organizations in designing organic farming and processing requirements according to their specific climatic and regional conditions. The US, EEC and Japan markets comprise the largest share in global trade on organic agricultural products. These major markets have been regulated by legislation enacted by respective countries that conform to guidelines laid under IFOAM Basic Standards. Therefore, it is essential to meet requirements of the US Organic Food Production Act, 1990, EEC Regulations on Organic Agriculture, 1991 and the Japanese Organic products Standard/Certification Act, 1993 in order to have global access to organic products. The common objective of these regulations are consumer protection by legally defining and making public the basis of organic food label claims. While defining the conditions of production in organic agriculture, the regulation considers the norms established in a country or region by taking into account the particular landscape climate and cultivation of the specific region. For example, the Soil Association of UK has defined the standards for organic food and farming which is implemented.

1.5 IPM strategy an eco-friendly approach for managing the pest in vegetables

The farm-gate survey conducted on pesticide residue shows that about 50% contaminations of vegetables is in UP and Kerala. Hence, it is need to adopt other alternate methods of pest management with emphasis on botanical formulations, biological, chemical and adapting suitable cultural practices. The nursery management is also helpful, particularly to curb the carry over of active and inactive stages of insects and viruses to the main field. Brinjal seed treatment with imidacloprid @ 2.5 g/kg reduces the little leaf and leafhopper incidence. Intercropping of brinjal (2 rows) with coriander (one row) or fennel (1 row) minimizes the incidence of brinjal shoot and fruit borer. For fruit and shoot borer, integrated pest management technique has been found highly effective. This includes installation of sex pheromone baited plastic funnel traps at 10m distances just above canopy from 20 days after planting followed by weekly removal of infected shoots along with larvae and periodical cleaning of dry leaves and old infested fruits.

In tomato seed treatment with imidacloprid, netting of nursery, spray 4% NSKE minimizes whitefly population, TLCV and leafminor incidence. Single spray of confidor @ 0.3 ml/lit about 40 days before fruiting in the main field is suggested. Planting of 2 rows of marigold as trap crop with every 14 rows of tomato along with release of *Trichogramma brassiliensis* @ 50,000 parasitized eggs/ha at flowering stage and spray of Ha—NPV @ 350 LE/ha twice after flowering can manage the fruit borer satisfactorily. Use of synthetic pyrethroids and endosulfan alternatively with NSKE (4%) is also effective against fruit borer. Seed treatment with imidacloprid @ 2.5 g/ka is effective against leafhopper in okra during early vegetative stage. Further, NSKE (3%) spray with sticker @ 0.1% takes care of pest. Spraying of endosulfan and neem formulation at their half doses at flowering stage can reduce borer infestation. Okra sown during 2nd week of June minimizes the incidence of both the pests. Plant growth in chilli is retarded during initial stage due to thrips and mites complex. Seed treatment with imidacloprid is effective against these pests. Alternate spray of dicofol @ 2.75 ml/lit and wettable sulphur @ 2 g/lit is advised against mite. Summer ploughing and crop rotation have been found beneficial for fruit fly management. Spraying of carbaryl 50 WP @ 2 g/lit at cotyledon stage controls red pumpkin beetles. Management of fruit fly during fruiting stage in cucurbitaceous crops through bait trap or bait spray is effective. Food bait consisting of rotten banana pulp (1000 g) + yeast (10 g) + malathion (10 ml) + citric acid (3 g) may be used in flat earthen pot to attract and kill the adult flies. Smashed pumpkin pulp may also be used instead of banana pulp. Bait spray containing 20 ml malathion (50 EC) + 20 litres water + 500 g molasses attract adult flies and control the population. The spraying is to be done in one plant per 10 m^2 area and rest of the plants remain free from insecticides.

Diamond back moth population in cabbage can be managed by growing paired mustard rows with every 25 cabbage rows. The first rows of mustard should be sown 15 days before and second 25 days after cabbage planting. Dichlorovos 0.1% is sprayed at weekly interval on the mustard to kill the trapped DBM larvae. If required NSKE (5%) may be sprayed to control the larval population of DBM. Such practices conserve the population of *Cotesia plutellae,* a dominant natural enemy of DBM. The DBM in cabbage and fruit borer in tomato can be managed through intercropping of respective crops. Soil solarization with transparent polyethylene sheet during summer is most ideal to suppress the soil borne pathogens in nursery. Seed treatment with *Trichoderma* spp. 5-10 g/kg or soil treatment @ 10-25 g/m^2 nursery area is advised against damping off diseases. Soil solarization also reduced bacterial seedling blight in nursery of tomato and chilli.

1.6 Priorities on processing

India is the world's second largest producer of vegetables, but hardly 2% of the produce is processed. About 25-30 per cent of the production is lost due to lack of adequate processing and infrastructural facilities. The prominent processed items are canned and dehydrated vegetables. More recently produces like frozen vegetables, vegetable curries in restorable pouches, canned mushrooms and mushroom produces have been taken up for manufacture by the industry. The vegetable industry in India is highly decentralized. A large number of units are in the cottage/home scale and small-scale sector, having small capacities (upto 250 tonnes/annum). But big Indian and Multinational companies in the sector have large capacities in the range of 30 tonnes per hour or so. Since liberalization and withdrawal of excise duty on fruit and vegetable products there has been significant rise in the growth rate of industry. Number of fruit and vegetable processing units licensed under FPO exceeds 4920, inclusive of over 650 sweetened aerated water units. The estimated production of the industry stood at 9.60 lakh tonnes in 1996-97, whereas the installed capacity was 15.10 lakh tonnes.

With the advancement in technology and trade agreements like GATT (General Agreement on Trade and Tariff) under WTO (World Trade Organization) we are bound to adopt control measures like HACCP (Hazard Analysis and Critical Control Points) and SPS (Sanitary and Phytosanitary) procedures. The quality standards of preserved vegetables have been upgraded in many developed countries. There is an all round realization at every level that quality production of international standard is the watchword for survival now. Consequently, the processor, small or big, have started taking steps to improve quality right from the selection of the raw material and that of the ingredients. Side by side steps have also been taken up by them to upgrade the hygienic and sanitary conditions of the workers, plant and machinery so as to ensure quality of the finished products. There is a much greater emphasis now to invest in research and development and also in product innovation. The Agricultural and Processed Food Products Export Development Authority (APEDA) gives subsidy to upgrade the laboratory facilities. It has also brought out a book entitled "ISO 9000 on The Food Industry A Practical Guide". It deals with the entire spectrum of quality control, e.g. documentation, installation, certification of quality management system, etc.

1.7 Irradiation of vegetables – A novel approach for processing

Food irradiation is a processing technique with similar results to freezing or pasteurization. The food is exposed to doses of ionizing energy or radiation. At low doses, irradiation extends a products shelf life. At higher doses, this process kills insects, moulds, bacteria and other potentially harmful microorganisms. Considerable scientific research over past five decades revealed that food irradiation is a safe and effective form of processing. Food irradiation has been approved in 40 countries includes the United States, China, France and Holland. The World Health Organization (WHO), the American Association and the Scientific Committee of the European Union are internationally recognized bodies which support food irradiation.

In this process, food is exposed to ionizing radiation, either from gamma rays or an electron beam. Gamma rays are a form of radiation that shares some characteristics with microwaves, but in a more concentrated form. The rays pass through the food just like microwaves in a microwave oven, but the food doesn't heat up. Exposure to gamma rays does not make food radioactive. Electron beams are formed from electricity and do not need radioactive material to work. In both the cases, organisms that are responsible for spoiling foods such as insects, moulds and bacteria are killed. However, food irradiation cannot kill viruses. The most irradiation research has concentrated on tomatoes, potatoes, onions, garlic, and mushrooms. In tomatoes, the major aim is to suppress the decay mould *Alternaria.* In potatoes, onion and garlic irradiation is meant to inhibit sprouting, whereas in mushrooms irradiation is supposed to inhibit growth and openings of caps. For suppressing the rotting in tomato caused by *Alternaria* it requires more than 3 k Gy, while many varieties can tolerate only 1 to 1.5 k Gy. For inhibiting sprouting in potatoes, irradiation with 0.085 and 0.15 k Gy is recommended. Mushrooms also tolerate irradiation quite well. Canadian research indicates that 1 k Gy inhibits the growth of cultivated mushrooms and increased storage life.

1.8 The benefits of food irradiation

Some of the benefits of this processing technique include:

- Extended shelf life of some products.

- Reduction of spoilage.
- Reduced risk of food-borne diseases caused by micro-organisms, such as *Campylobacteria* and *Listeria* (especially in meat, poultry and fish).
- Lessened need for pesticides.
- Lessened need for some additives, such as preservatives and antioxidants.
- Lower risk of importing or exporting insect pests hidden inside food products.
- Reduced need for chemical treatments, such as those used to kill bacteria found in some spices.
- As an alternative to current treatment for disinfecting imported fruits, grains and vegetables, which use ozone-depleting gas.
- Reduced sprouting in potatoes and onions.

Irradiation technique has immense potential in prolonging the storage life of several vegetables. However, concerted research is required on irradiation sensitivity of fresh cut vegetables and its relationship with endogenous antioxidant capacity or phenolics content of the vegetables.

1.9 Using the nanotechnology for packaging

The increasing demand for premium packaging and packing system that offers ready added value needs concerted research efforts. Further, increasing health consciousness among consumers, food packaging is also more inclined towards aspects such as ensuring food safety through control of the environment within the packaging and minimizing damage resulting from microbial attacks. Recently, the foundation for Scientific and Industrial Research at the Norwegian Institute of Technology (SINTEF) is attempting to use nanotechnology to develop plastic film that provide barrier protection by preventing gases such as oxygen and ethylene from damaging the food content.

1.10 Export of vegetables

India's export of fresh vegetables in 2003-04 registered a steep growth of 48.19 per cent over the previous year when the same reached a level of Rs. 27.39 crores as against Rs. 11.82 crores. This was followed by onions (97.86%) and peas (11.36%). Onions, as may be seen, continues to be the largest item of export. It alone accounted for a share of 76.18 per cent in the total export of vegetables from the country in 2003-04. Other vegetables , next in importance, comprised cucumbers and gherkins, mixed vegetables, (potatoes and mushrooms).

1.11 Recent export promotion measures

To boost exports of vegetables, several initiatives have recently been taken up both by the Government and the trade and industry. Some of the major ones are given below:

(i) New Foreign Trade Policy (FTP): Under the new FTP (2004-09) announced by the Ministry of Commerce and Industry, Government of India, have been given to boost agri-exports. These inter alia include (a) duty-free import of capital goods under the Export Promotion Capital Goods (EPCG) scheme, (b) Duty credit scrip equivalent to 5 per cent of the f.o.b. value of exports and the (c) Launching of Vishesh Krishi Upaj Yojana which is aimed at promoting agri-exports, viz., vegetables, fruits, flowers, minor forest produce, etc. (ii) Setting up of JV, Bharti Enterprises Ltd. has entered in the field of agri-exports through setting up of a 50:50 joint venture with a UK based company, Rothchild Group, ELRO Holdings Ltd. The JV "Field Fresh Foods" will have an investment of Rs. 250 crores.

The World Trade Organization (WTO) and General Agreement on Tariffs and Trade (GATT) have opened the new opportunities for the country to compete at the global level for exporting fresh and processed vegetables. Accordingly, India is emerging as one of the

Table 1.1 : Variety wise India's exports of fresh vegetables in the years 2004-05 and 2005-06

Hs Code	Variety	2004-05	2005-06	% change in 2004-05 over 2005-06
07.03.1001	Onions (fresh or chilled)	361.80	715.87	97.86
07.07.0000	Cucumbers and gherkins (fresh or chilled)	47.40	39.94	(-) 15.74
07.09.9003	Mixed vegetables	47.97	37.24	(-) 22.37
07.09.9000	Potatoes (fresh or chilled)	11.82	27.29	134.11
07.09.5100	Mushrooms (fresh or chilled)	51.77	23.53	(-) 54.55
07.09.6001	Green chilli	28.45	8.99	(-) 68.40
07.08.1000	Peas shelled or unshelled (fresh or chilled)	7.66	8.53	11.36
07.02.0000	Tomatoes (fresh or chilled) other than seeds	11.85	65.85	(-) 42.19
Total (All India)		**642.78**	**939.70**	**46.19**

major exporting countries for a number of vegetable crops. During the year 2001-02, an amount of Rs. 582.31 crores have been realized with the export of total fresh vegetables. This has been possible and the quality of vegetables has been improved with evolution and adaptation of new varieties/hybrids, integrated nutrient management, improved methods of crop production, integrated disease and pest management technologies and post harvest management.

1.12 Seed-a fulcrum of all multiplication programmes

Seed is the most important input component for agricultural production. In the significant advances that India made in agriculture in last four decades, the role of the seed sector has been substantial. The expansion of seed industry has occurred in parallel with growth in agricultural productivity. The Indian seed industry is currently valued around RS. 2500 crores ($500 million) and its proposed estimates for the year 2002 was 3750 crores ($750 million). There are about 150 organized seed companies in India today, besides the Public sector component which comprises two central corporations viz., National Seeds Corporation (NSC) and State Farms Corporation of India (SFCI) and 13 State Seed Corporations. Several companies have Government of India (DSIR) recognized research and development departments and have produced and released a large number of varieties and hybrids in several crops. Private R & D's real investment in research has quadrupled between 1986 and 1998. Subsidiaries and joint ventures with multinational companies indicate that the amount spent on R & D ranged from 0.78 per cent (0.49 crore) to 15.08% (22.62 crores) (Companies Annual Report 1998-99).

The seed components in 1990-91 comprised of proprietary hybrids, public bred hybrids and open pollinated varieties. In terms of quantity and value, seed of open pollinated varieties was the largest followed in order by public hybrids and proprietary hybrids. Seeds of proprietary hybrids formed a significant portion of the total seed market. In terms of quantity, although proprietary hybrids had only 32.34 per cent share of the market, in terms of value the share was 76%. In contrast to the 1991 scenario, the estimates in 1998-99 with proprietary hybrids growing at the expense of public hybrids. The area planted to bring seed increased by 3 per cent over that of 1990-91 and the market size expanded significantly in terms of quantity and value. The total market for purchased seed was 8.64 lakh tonnes valued at Rs. 2249 crores. The component of proprietary hybrid seed was estimated to be around 51,314 tonnes in 1998-99 as against 19,300 tonnes in 1990-91 and valued at Rs. 636 crores as against Rs. 95 crores, respectively (Table 1.2).

Table 1.2 : Indian seed market (1990-91 and 1998-99)

Item	1990-91	1998-99
Proprietary hybrids	95.00	636.00
Public bred hybrids	124.65	245.00
Varieties	460.15	1368.55
Grand total	679.80	2249.55

Source: (a) The Market for seed in India: A syndicated report based on secondary research (1991) by Francis Kanoi. (b) Marketing Database, bench marking the seed market : Mahyco, 1999.

The availability of quality seed is *sine qua non* for enhancing the vegetable production in the country. As the cost of seed is only a tiny fraction of total cost of inputs involved in growing the vegetable crops, providing good quality seeds is one of the most important and easiest means to accelerate the productivity of vegetables in the country. Total requirement of seeds in our country is 48,000 tonnes and less than one-third of this is met by public and private sector undertakings.

In order to organize successful v\breeder, foundation and certified seed production programmes, research back up in terms of development and availability of information on appropriate seed production methods, techniques of seed quality maintenance and improvement, storage and seed health care is most vital and essential. Realizing the importance of seed in increasing vegetable production, the Government of India launched a National Seed Project (NSP) in October 1976 with World Bank Assistance. Since April 1980, the Council took the continuance of the project as its plan activity during VI and VII plan periods. A centrally sponsored scheme on vegetable breeder seed production having 7 centres which terminated in March 1985 was included in ICAR VII plan of NSP. The independent breeder seed production programme of vegetables was initiated in VIII plan.

Under NSP, ICAR is guiding, coordinating and promoting seed technology research and breeder seed production in its Institutes and State Agricultural Universities. In addition to the responsibility of production of nucleus and breeder seed of varieties and parental lines of hybrids, the production programmes are duly supported by nucleus seed production and researches on various aspects of seed technology like isolation requirement, genetic purity tests, seed physiology and seed health care, processing, packaging, storage etc.

A separate NSP on vegetables was approved and initiated by IIVR, Varanasi in 1994, Under this for breeder seed production, ICAR has set up 4 units at ICAR Research Institutes and 10 units in the State Agricultural Universities. In the Divisional concept of the Council, now there are 14 breeder seed production units dealing with vegetables (excluding potato) are being looked after by Indian Institute of Vegetable Research, Varanasi of Horticulture Division of the Council. The breeder seed production under the programme as per the indent received is given in the Table 1.3.

Table 1·3 : The breeder seed produced during the period from 1997-98 to 2001-02

Year	Indent (kg)	Production (kg)
1997-1998	71603.00	71303.00
1998-1999	43874.47	47974.35
1999-2000	46647.73	51103.25
2000-2001	55799.65	60020.65
2001-2002	37843.40	45102.86

1.13 Export of vegetable seeds

Although India is second largest vegetable producing country in the world, it has 8th rank among the sixteen countries listed in seed trade the world over. In recent years, a lot of emphasis is being given to improve the quality of seed and planting materials in view of globalization of the world trade which had opened the ample opportunities for the export of quality seed. Enjoying the diversity of agroclimatic conditions, strong seed production infrastructure and market opportunity, India holds significant for export of seeds. The export of fruits and vegetable seed during 2001-02 was 6188.67 metric tonnes amounting to Rs. 6457.89 lakhs (Table 1.4). For fruits and vegetable seeds USA, Japan, Netherlands, Italy and Pakistan were major markets for the year 2001-02.

Table 1.4 : Export of seeds of major vegetables (Rs.)

Vegetable	Rupees		
	1998-99	**1999-00**	**2000-01**
Potato	39584130	15398192	14387935
Cabbage	5657697	484910	23494487
Cauliflower	3194145	591608	3511670
Onion	5660434	16141209	11384220
Pea	3845670	826291	160993
Radish	178650		69245
Tomato	162452647	110806580	87516676
Others	290601538	189881409	248583846

Source : APEDA, New Delhi

1.14 New seed policy

The new policy on seed development greatly liberalized import of vegetables and flower seeds, in general and seeds of other commodities in a restricted manner and also encourages multinational seed companies to enter the seed business. More than 24 companies initiated research and development activities and have made substantial commitments for investment on research and development in response to this policy initiative. Further, a draft Seed Act 2001 is being finalized on the basis of the recommendations of Seed Policy Group. It would replace the existing Seed Act 1966 and Seed (Control) order of 1983. The proposed legislation features establishment of National Seed Board (NSB) and compulsory registration of any seed with the board before sowing planting could be done for commercial purposes.

1.15 Fastening the market intelligence and communication network

The availability of prompt and reliable market information about the market arrival and prices for different commodities considerably improves the decision-making capacity of the farmers. The information collected by the conventional method is stale and does not back the farmer/producer about the prevailing prices in other markets where he can dispose off his produce in the nearest markets even at unfavorable prices. A need has been felt to improve the existing marketing information system by linking all the regulated markets in the country with computer to control electronic communications (e-mail with modem) with free access to information. The important communication system will enable the producers to know about market where he can dispose off his produce profitably. The producers and consumers can also derive maximum advantages out of their purchases at low communication cost. Installation of computers at different levels is required for better research management for the benefit of different market users through NIC NET facilities. The State Government should also make efforts to provide for such improved communication system at various levels in the states.

National Horticultural Board has a network for collection of market information for fruits and vegetables in 33 important markets. The information is collected, compiled and disseminated through radio, newspapers and monthly bulletins. All the important markets in the country need to be linked with each other through computer network so that they are in touch with each other regarding trends in arrivals, prices, dispatches, etc.

1.16 Constraints in vegetable production

(1) *Lack of planning in production*

Reliable statistics on area, production and productivity are not available which posses hindrance for planning and execution of development programmes. Technology transfer, information network for marketing and training of farmers are not given adequate attention at government level. For this reason, unplanned production is a feature in India. In our country, vegetable production is still being treated as a low priority area.

(2) *Non availability of seeds of improved varieties*

Vegetable seed production system and network at government level is ill-developed which hardly contributes 5-10% of the total annual requirement of about 26 thousand tonnes. As many as 600 improved varieties have been developed/identified at national/state level, but hardly 30% of the area is covered by these varieties. Quality seeds alone can bring about at least 25-30 per cent increase in productivity level.

(3) *High cost of basic production elements*

Vegetable cultivation is basically an input intensive venture. Price hike of fertilizers restrains the farmers from applying these inputs in proper way, particularly when the profitability is not assured. The cost of hybrid seeds is extremely high, thus small and marginal farmers often cannot afford to purchase such seeds.

(4) *Inadequate plant protection measures and non-availability of resistant varieties*

Most of the vegetable crops are highly susceptible to diseases and as yet there is no effective control measures to check diseases like leaf curl in tomato and chillies, blights in tomato, wilts in tomato and brinjal, etc. Similarly, fruit and shoot borers in brinjal, diamond back moth in cabbage, fruit fly in cucurbits, thrips in onion and nematodes in different vegetable crops are some major enemies. There is lack of resistant varieties to such biotic stresses on production.

(5) *Weak marketing facilities*

There is absolute lack of linkage between production and marketing. Stronghold of middlemen in marketing of vegetables exists due to inadequate cooperative system. For the reason, prices often fall very low and farmers are compelled for distress sale which reduces the incentive for higher production.

(6) *Transportation limits*

In India, cold chain facility has not so far been developed for both rail and road transport. Vegetables transported by train are often overstocked, long awaited and even decayed. Road transport is now more reliable than railway transport but it involves higher cost. Even in road transport, about 30 to 40 per cent produce gets deteriorated due to lack suitable packaging materials and pretreatment facilities. Concept of rotation freezers and refrigerated vans is still remote in this country.

(7) *Post-harvest losses*

In our country, about 30-40 per cent produce becomes unfit for consumption before it reaches the consumer. There is no suitable infrastructural facilities for preventing such huge post-harvest losses. Most of the cold stores have been developed for storing potato, Moreover, hardly 0.5 per cent produce is processed to make durable products.

(8) *Inadequate irrigation facilities*

In the vast areas of the country, proper irrigation facility has not yet been provided. Vegetable crops are not being grown successfully in such areas.

(9) *Abiotic stresses*

Vegetable crops are sensitive to soil salinity and alkalinity and therefore, they cannot be grown in such areas without soil amendment and/or developing salt or alkali-tolerant varieties.

(10) *Unstable vegetable fields*

Due to urbanisation and industrialisation, the vast productive vegetable fields around suburbs have been requisitioned. Local laws or regulation to prevent vegetable field from uncontrolled requisition and protect the basic area for vegetable cultivation is inadequate.

Future thrust area

1. Working out the area and production of the vegetable crops for proper planning good budgetary support is essentially needed to tackle various problems of vegetable production and consumption.
2. Launching of adequately funded food losses prevention programme to provide infrastructures for well-coordinated research and development work including processing.
3. Immediate farming of a national seed policy to ensure successful, economic and dependable supply of seeds of improved varieties.
4. Development of hybrid seed production village by imparting training so that the farmers are trained in taking up the hybrid seed production programmes which will not generate more employment in rural sector but will also reduce dependency on seed import.

5. Development of varieties combining higher yield potential and yield dependability with disease and pest resistance.
6. Improvement of transfer of technology at farmers level and in this approach, expert personnel of vegetable crops must be involved in extension services.
7. Exploitation of large new areas for vegetable cultivation and in this regard, vegetable production must be promoted in the tribal areas.
8. Streamlining of export oriented production through small and marginal farmers and more attention should be paid to off season production.
9. Development of cooperative marketing system to ensure more profit to the growers.
10. Promotion of research on rainfed and riverbed vegetable production.
11. Generation of drip irrigation facilities and other irrigation resources in vegetable cultivation.
12. Quick development and sustainable increase of protected cultivation by adopting polyhouse (polyethylene booth) and polyethylene tunnel technology.
13. Rapid extension of plastic mulch technology to exploit solar decomposed and bio-composed plastic film in order to prevent soil pollution.
14. Rapid development and commercialization of F_1 hybrids.
15. Strengthening of research on biotechnology, such as, genetic engineering, genetic transformation, tissue culture, protoplast fusion, etc.
16. Drawing of regular schemes by the banks, NABARD (National Bank for Agriculture and Rural Development) and rural credit societies with close coordination of the Central and State Governments to provide loans to protect the small and marginal farmers from being exploited by the money lenders.
17. Introduction of specific crop insurance scheme by the Government to safeguard the vegetable growers as vegetable crops are sensitive to climatic trickery and the growers sometimes lose everything whatever they have invested.

The objective type of questions on general information on vegetables are given below :

1. Fruits of vegetable marrow are ready for first picking after *(ARS 2003)*

   (a) 40-50 days (b) 60-80 days
   (c) 80-100 days (d) None of these

2. Fungicide used for controlling powdery mildew disease of pea is *(ARS 2003)*
   (a) Captan (b) Indofil M-45
   (c) Karathane (d) None of these
3. Powdery mildew disease is a serious problem of *(ARS 2003)*
   (a) Capsicum (b) Okra
   (c) Cucurbits (d) Brinjal
4. Pusa Chikni is a variety of *(ARS 2003)*
   (a) Sponge gourd (b) Ridge gourd
   (c) Bottle gourd (d) Bitter gourd
5. .... variety of *Brassica pekinensis* which is a useful leafy vegetable for plains and hills of North India. *(ARS 2003)*
   (a) Chinese (b) American
   (c) Canadian (d) Japanese
6. Type of vegetable gardening followed on the Dal Lake of Kashmir Valley is *(ARS 2003)*
   (a) Floating garden (b) Hydroponics
   (c) Kitchen garden (d) None of these
7. Types of vegetable gardens according to the purpose for which they have been developed are *(ARS 2003)*
   (a) Two (b) Four
   (c) Six (d) None of these
8. Brinjal variety resistant to the attack of fruit borer is *(ARS 2003)*
   (a) Punjab Neelum (b) Punjab Barsati
   (c) Jamuni Gola (d) None of these
9. Varsha Upahar is a variety of *(ARS 2003)*
   (a) Okra (b) Cabbage
   (c) Chilli (d) Cowpea
10. Vegetables are packed in cans and processed by heat in *(ARS 2003)*
    (a) Retourt (b) Water bath
    (c) Sun (d) Any other means
11. For getting higher number of female flowers in bottle gourd, spray the plants at 2-4 leaf stage with boron in a concentration of *(ARS 2003)*
    (a) 20 ppm (b) 10 ppm
    (c) 3 ppm (d) None of these
12. For growing carrot, radish and turnip for table purpose, sowing is done at a spacing of *(ARS 2003)*
    (a) 45 x 7.5 cm (b) 30 x 15 cm
    (c) 45 x 7.5 cm (d) None of these

13. For induction of greater number of female flowers in watermelon, spray seedlings at 2-4 leaf stage with 2,4,5-tri iodobenzoic acid in the concentration of *(ARS 2003)*
    (a) 10-15 ppm (b) 25-50 ppm
    (c) 50-100 ppm (d) None of these
14. Vegetables are rich source of *(ARS 2003)*
    (a) Carbohydrates (b) Proteins
    (c) Vitamins (d) None of these
15. Vegetables are subjected to drying after *(ARS 2003)*
    (a) Blanching (b) Without blanching
    (c) Sulphuring (d) Any other means
16. Chemical used for controlling root-knot nematodes in brinjal is *(ARS 2003)*
    (a) Aldrin (b) Nemagon
    (c) Chloropyriophos (d) None of these
17. *Chenopodiaceae* is the family of *(ARS 2003)*
    (a) Potato (b) Sweet Potato
    (c) Okra (d) Spinach
18. Which of the following is an often cross pollinated vegetable crop? *(ARS 2003)*
    (a) Tomato (b) Bhindi
    (c) Pea (d) Onion
19. Which of the following is a fruit vegetable crop? *(ARS 2003)*
    (a) Sweet potato (b) Okra
    (c) Potato (d) Spinach
20. Which of the following is recommended isolation distance for certified seed production of cabbage? *(ARS 2003)*
    (a) 200 m (b) 400 m
    (c) 800 m (d) 1000 m
21. Which of the following is a self pollinated vegetable? *(ARS 2003)*
    (a) Bottle gourd (b) Brinjal
    (c) Radish (d) Cauliflower
22. Which one is not correctly matched? *(ARS 2003)*
    (a) Ginger- Rhizome (b) Onion- Bulb
    (c) Potato- Tuber (d) Sweet potato- Root
23. Which of the following is used to check sprouting of onion bulbs under storage? *(ARS 2003)*
    (a) N.A.A. (b) M.H.
    (c) G.A. (d) P.C.P.A.

24. Type of flowers in ridge gourd is (ARS 2003)
    (a) Andromonoecious (b) Hermaphrodite
    (c) Gynoecious (d) All of the above
25. Cracking of tomato fruit is due to (ARS 2003)
    (a) Restricted watering
    (b) A period of drought followed by sudden heavy watering
    (c) Over watering (d) None of these
26. Best temperature for storing tomato fruits is (ARS 2003)
    (a) 4.5°C (b) 20 to 25°C
    (c) 12 to 15°C (d) 15 to 20°C
27. Male sterility has been commercially exploited for the hybrid seed production in (ARS 2003)
    (a) Cucumber (b) Muskmelon
    (c) Water melon (d) None of these
28. Pollination in beet root is (ARS 2003)
    (a) By insect (b) By wind
    (c) By water (d) By all these
29. Wind pollination plays an important role in seed production of (JRF, 2005)
    (a) Radish (b) Turnip
    (c) Garden beet (d) None of these
30. Yellow coloured vegetables are rich source of (ARS 2003)
    (a) Vitamin E (b) Vitamin C
    (c) Vitamin A (d) Vitamin B
31. First intergenic cross between radish and cabbage was made by (ARS 2003)
    (a) Jennings (b) Karpenchiko
    (c) Hull (d) Richy
32. Flower initiation and higher yields may be expected with the treatment of onion bulbs with (ARS 2003)
    (a) NAA (b) IAA
    (c) IBA (d) GA_3
33. Flower of brinjal may be (ARS 2003)
    (a) Long styled (b) Medium styled
    (c) Pseudo short styled (d) All of the above
34. For a balanced diet, the dieticians have recommended daily consumption of leafy vegetables to the tune of (ARS 2003)
    (a) 50 g (b) 116 g
    (c) 250 g (d) None of these

35. The All Indian Coordinated Vegetable Improvement Project was started in *(ARS 2003)*
    (a) 1968 (b) 1970
    (c) 1984 (d) 1987

36. Sree Prakash is the variety of *(ARS 2003)*
    (a) Cassava (b) Tomato
    (c) Lettuce (d) Celery

37. Heterotic progeny from a cross between two inbreds is known as *(ARS 2003)*
    (a) Single cross hybrid (b) Double cross hybrid
    (c) Three way cross hybrid (d) multiple cross hybrid

38. Mendel's results were first published in *(ARS 2003)*
    (a) English (b) French
    (c) German (d) Russian

39. In garden pea, Mendel studied the dominant and recessive behaviour of *(ARS 2003)*
    (a) 5 characters (b) 6 characters
    (c) 7 characters (d) 8 characters

40. Basic chromosome number of spinach beet is *(ARS 2003)*
    (a) 12 (b) 9
    (c) 6 (d) 20

41. Depending upon the type of seed, some seed borne diseases are effectively controlled by soaking the seed for 20-30 minutes in hot water at a temperature of *(ARS 2003)*
    (a) 30°C (b) 50°C
    (c) 70°C (d) None of these

42. Indian Institute of Vegetable Research is situated at *(ARS 2003)*
    (a) Bhopal (b) New Delhi
    (c) Varanasi (d) Ludhiana

43. The most ancient type of garden is *(ARS, 1999, JRF, 2005)*
    (a) Kitchen garden (b) Market garden
    (c) Vegetable garden (d) None of these

44. The most serious viral disease of tomato in the rainy and autumn seasons is *(ARS 2003)*
    (a) TMV (b) CMV
    (c) Leaf curl
    (d) Tomato spotted wilt virus

45. The most troublesome insect pest of lettuce is *(ARS 2003)*
    (a) Leaf miner (b) Aphid
    (c) Painted bug (d) None of these

46. How one can protect the stem end of radish root from becoming green? *(ARS 2003)*
(a) By thinning (b) By staking
(c) By earthing (d) By all of these

47. Which of the following vegetables is rich source of vitamin C? *(ARS 1992)*
(a) Potato (b) Sweet potato
(c) Garlic (d) Green chilli

48. Leaf curl virus in nature is spread by *(ARS 1992)*
(a) Aphid (b) White fly
(c) Leaf hopper (d) None of these

49. Watermelon belongs to the species *(Ph.D. VGS, 2006)*
(a) *vulgaris* (b) *sativus*
(c) *lanatus* (d) None of these

50. Central Institute of Horticultural Research for Northern plains is situated at *(UP PCS (pre) 1994)*
(a) Patna (b) Bangalore
(c) Lucknow (d) Kolkatta

51. Which of the following statements is correct? *(Ph.D. VGS 2006)*
(a) Kufri Badshah is late blight resistant variety of potato
(b) Pant Bahar is a variety of brinjal
(c) Pant-C is a variety of tomato
(d) Pusa sawni is a variety of onion

52. Which one is a monocot vegetable? *(Ph.D. 2006)*
(a) Carrot (b) Fenugreek
(c) Parwal (d) Colocasia

53. Which one is an insect trap crop? *(Ph.D. 2006)*
(a) Bhindi (b) Potato
(c) Onion (d) Paddy

54. Which of the following vegetables does not belong to family *Alliaceae*? *(Ph.D. 2006)*
(a) Onion (b) Garlic
(c) Asparagus (d) Leek

55. Arka Gaurav is a variety of *(ARS (Pre) 2003)*
(a) Tomato (b) Brinjal
(c) Capsicum (d) Onion

56. In which of the following vegetables climatic phenomenon is observed ? *(Ph.D. VGS, 2006)*
(a) Peas (b) Okra
(c) Watermelon (d) Tomato

57. In which of the following vegetables heterostyly phenomenon takes place ? *(Ph.D. VGS, 2006)*
    (a) Onion (b) Brinjal
    (c) Tomato (d) Garlic

58. Which of the following term is appropriate for dry petha sweet? *(Ph.D. VGS, 2006)*
    (a) Fruit candy (b) Preserve
    (c) Candied (d) Morabba

59. CFTRI is situated at *(UPPCS (Pre) 1996, Ph.D. VGS 2004)*
    (a) New Delhi (b) Hyderabad
    (c) Bangalore (d) Mysore

60. In which of the following methods grains of vegetable peas can be kept for longest period? *(Ph.D. VGS, 2006)*
    (a) Cold storage (b) Deep freeze
    (c) Ordinary storage (d) None of these

61. Mulching in brinjal is recommended with *(Ph.D. VGS, 2006)*
    (a) Black polythene film (b) White polythene film
    (c) Straw (d) Sand

62. The best suitaed pH range for brinjal cultivation is *(RAS (Pre) 2003)*
    (a) 4.5 to 5.5 (b) 5.5 to 6.0
    (c) 5.5 to 3.6 (d) 6.0 to 7.0

63. Greening of potato results in *(UP PCS (Pre) 1996)*
    (a) Increase in nutritional quality
    (b) Decrease in nutritional quality
    (c) Increase in disease resistance
    (d) Decrease in disease resistance

64. Optimum temperature for fruit setting in cucumber and muskmelon ranges between *(Ph.D. VGS, 2006)*
    (a) 55°F and 65°F (b) 70°F and 80°F
    (c) 40°F and 50°F (d) None of these

65. "Seed plot technique" is followed for seed production in *(UP PCS (Pre) 1996)*
    (a) Wheat (b) Paddy
    (c) Potato (d) Bajra

66. Puffiness is a physiological disorder in *(RAS (Pre) 2003)*
    (a) Brinjal (b) Chilli
    (c) Okra (d) Tomato

67. Leghaemoglobin is a red plant pigment in *(Ph.D. VGS, 2006)*
    (a) Red chillies
    (b) Root of carrot
    (c) Ripe fruits of tomato
    (d) Root nodules of leguminous plants
68. "Seed plot technique" in potato was proposed by *(HP. PCS 2003)*
    (a) Dr. J. S. Grewal
    (b) Dr. Mukhtar Singh
    (c) Dr. B. B. Nagaich
    (d) Dr. Pushkar Nath
69. Function of tuber is *(Ph.D. VGS, 2006)*
    (a) To store reserve food
    (b) Propagation
    (c) Both
    (d) None of these
70. Turmeric is propagated by *(Ph.D. VGS, 2006)*
    (a) Rhizomes
    (b) Corms
    (c) Tubers
    (d) Stolons
71. Potato, tomato, brinjal and chillies belong to the family *(Ph.D. VGS, 2006)*
    (a) *Cucurbitaceae*
    (b) *Solanaceae*
    (c) *Papilionaceae*
    (d) *Cruciferae*
72. Pusa Ruby, Pusa Early Dwarf, Sioux, Marglobe, Best of All, SI-120 and Roma are varieties of *(Ph.D. VGS, 2006)*
    (a) Brinjal
    (b) Sweet pepper
    (c) Tomato
    (d) Cucumber
73. Pusa Purple Long, Pusa Purple Round, Benares Giant, Black Beauty, Mukta Keshi, Surti Gota are varieties of *(Ph.D. VGS, 2006)*
    (a) Potato
    (b) Sweet pepper
    (c) Tomato
    (d) Brinjal
74. California Wonder and Yellow Wonder are varieties of *(Ph.D. VGS, 2006)*
    (a) Sweet pepper
    (b) Brinjal
    (c) Tomato
    (d) Cucumber
75. Pusa Meghali is an improved variety of *(UP PCS (Pre) 2003)*
    (a) Carrot
    (b) Radish
    (c) Beet root
    (d) Turnip
76. Browning in cauliflower is due to *(UP PCS (Pre) 2003)*
    (a) Boron deficiency
    (b) Nitrogen deficiency
    (c) Excess of boron
    (d) Molybdenum deficiency
77. Crimson Globe and Detroit Dark Red are varieties of *(Ph.D. VGS, 2006)*
    (a) Beet
    (b) Carrot
    (c) Turnip
    (d) Onion

78. Which of the following crop varieties is available for growing at three different times during an year? *(Ph.D. VGS, 2006)*
    (a) Cauliflower (b) Brussels sprout
    (c) Chinese cabbage (d) Cabbage

79. The isolation distance for producing certified seed of cauliflower is maintained at *(JRF, 1998, UPSC (Pre), 2000)*
    (a) 500 m (b) 1500 m
    (c) 1000 m (d) 2000 m

80. The isolation distance for producing certified seed of brinjal is maintained at *(Ph.D. VGS, 2006)*
    (a) 100 m (b) 200 m
    (c) 300 m (d) 400 m

81. The isolation distance for producing certified seed of chilli is maintained at *(Ph.D. VGS, 2006)*
    (a) 400 m (b) 300 m
    (c) 200 m (d) 100 m

82. The isolation distance for producing certified seed of tomato is maintained at *(Ph.D. VGS, 2006)*
    (a) 100 m (b) 75 m
    (c) 50 m (d) 25 m

83. Vegetable forcing refers to *(Ph.D. VGS, 2006)*
    (a) Production of off season vegetables
    (b) Soilless cultivation of vegetables
    (c) Growing of vegetables on roof top
    (d) Growing of vegetables in large containers

84. Late cauliflower varieties are available for consumption in *(ARS, 1992)*
    (a) Mid January to April (b) May to June
    (c) December to January (d) June to July

85. Arka Niketan is a variety of *(ARS, 1992)*
    (a) Onion (b) Garlic
    (c) Shallot (d) Welsh onion

86. *Brassica oleracea var. italica* is botanical name of *(ARS, 1992)*
    (a) Knol-khol (b) Kale
    (c) Brussels sprout (d) Broccoli

87. Warm season vegetable is *(ARS, 1992)*
    (a) Onion (b) Garlic
    (c) Sponge gourd (d) Radish

88. Pusa Anupam is a variety of *(ARS, 1992)*
    (a) Tomato (b) Brinjal
    (c) Chilli (d) Capsicum

89. Pusa Phalguni, Pusa Barsati and Pusa Dofasli are varieties of *(ARS, 1992)*
 (a) French bean (b) Cowpea
 (c) Pea (d) Mungbean
90. Pusa Parvati, Contender and Stringless are varieties of *(ARS, 1992)*
 (a) French bean (b) Cowpea
 (c) Cluster bean (d) Mungbean
91. Lincoln, Bonneville, Meteor, Early Badger and Arkel are varieties of *(ARS, 1992)*
 (a) Peas (b) Beans
 (c) Cowpea (d) Mungbean
92. Pusa Saffaid, Pusa Lal, Pusa Sunheri, Bhadrakali, Contennial and Housur red are varieties of *(ARS, 1992)*
 (a) Sweet potato (b) Potato
 (c) Turnip (d) Radish
93. All Green and Pusa Jyoti are varieties of *(ARS, 1992)*
 (a) Spinach (b) Palak
 (c) Fenugreek (d) Amaranth
94. Asparagus, rhubarb, globe artichoke, jerusalem artichoke are *(ARS, 1992)*
 (a) Annual vegetables (b) Perennial vegetables
 (c) Biennial vegetables (d) Dry vegetables
95. Pusa Mukta is an important variety of *(UP PCS (Pre), 2003)*
 (a) Cabbage (b) Cauliflower
 (c) Turnip (d) Radish
96. Pant Rituraj is a variety of *(UP PCS (Pre), 2000)*
 (a) Tomato (b) Cauliflower
 (c) Brinjal (d) Cabbage
97. Pusa Gaurav is an improved variety of *(UP PCS (Pre), 2004)*
 (a) Brinjal (b) Radish
 (c) Tomato (d) Chilli
98. Canning media for vegetable is *(UP PCS (Pre), 2004)*
 (a) Brine (b) Sugar
 (c) Brine and Sugar (d) None of the above
99. The crop produced maximum in the world is *(UP PCS (Pre), 2004, 2005)*
 (a) Mango (b) Tomato
 (c) Potato (d) Grape
100. "Arka Anupama" is a variety of *(UP PCS (Pre), 2005)*
 (a) Spinach (b) Amaranth
 (c) Beet root (d) Pea

101. "Azad Nutan" is a variety of *(UP PCS (Pre), 2005)*
(a) Bottle gourd (b) Bitter gourd
(c) Cucumber (d) Brinjal

102. The maximum vegetable producing country in the world is *(UP PCS (Pre), 2005)*
(a) China (b) Australia
(c) South Africa (d) None of the above

103. Maximum calcium is found in *(UP PCS (Pre), 2005)*
(a) Onion (b) Methi
(c) Cow pea (d) None of these

104. "Buck eye rot" is a disease of *(UP PCS, 2004 (Pre), 2005)*
(a) Sweet potato (b) Tomato fruits
(c) Pods of garden pea (d) Water chestnut

105. Which of the following is a stem vegetable *(UP PCS (Pre), 2005)*
(a) Carrot (b) Sweet potato
(c) Knol-khol (d) Radish

106. Chief Editor of Advances in Horticulture is *(ARS, 1999)*
(a) Dr. Kirti Singh (b) Dr. K. L. Chadda
(c) Dr. G. Kalloo (d) None of these

107. For proper growth and development of tomato and Sweet potato optimum temperature is *(ARS, 1999)*
(a) 20-25°C (b) 15-17°C
(c) 20-30°C (d) None of these

108. Significant contribution in spice improvement in India has been mediated by *(ARS, 1999)*
(a) Dr. G. Kalloo (b) Dr. Vishnu Swaroop
(c) Dr. K.V. Peter (d) None of these

109. Indian Institute of Spices Research is situated at *(ARS, 1999)*
(a) Cochin (b) Calicut
(c) Trivendrum (d) None of these

110. No. of seed companies in India are about *(ARS, 1999)*
(a) 400 (b) 500
(c) 600 (d) 700

111. During 2004 vegetable production in India was *(ARS, 1999)*
(a) 70 mt (b) 75 mt
(c) 85 mt (d) 95 mt

112. In India (2003-04) area under vegetable production is *(ARS, 1999)*
(a) 4-0 mha (b) 5.0 mha
(c) 6.2 mha (d) None of these

113. 'Hydroponics' term was given by (ARS, 1999)
(a) Marlobe (b) Jinks
(c) Gericke (d) None of these

114. Dimond black moth is the serious pest of (ARS, 1999)
(a) Cauliflower (b) Cabbage
(c) Tomato (d) None of these

115. CPRI at Shimla was established in (ARS, 1999)
(a) 1949 (b) 1950
(c) 1951 (d) 1947

116. Protein is found in abundance in (ARS, 1999)
(a) Bitter gourd (b) Cabbage
(c) Peas & Beans (d) None of these

117. Floating garden is generally found in (ARS, 1999)
(a) Delhi (b) Bangalore
(c) Pune (d) Kashmir

118. Arka Mohini is the variety of (ARS, 1999)
(a) Pea (b) Potato
(c) Chilli (d) None of these

119. Sugar Baby is the variety of (ARS, 1999)
(a) Muskmelon (b) Pumpkin
(c) Water melon (d) None of these

120. Pusa Purple Long is the variety of (ARS, 1999)
(a) Brinjal (b) Tomato
(c) Potato (d) None of these

121. Arka Niketan is a variety of (ARS, 1999)
(a) Onion (b) Brinjal
(c) Cauliflower (d) None of these

122. Botanically edible portion of lotus (Kamal kakri) is (ARS, 1999)
(a) Shoot (b) Root
(c) Crown (d) None of these

123. White rust (Blisters) of crucifers is caused by (ARS, 1999)
(a) *Steptomyces scabies* (b) *Albugo candida*
(c) *Olpidium branical* (d) *Cereosporella branicae*

124. Bathuwa is botanically called (ARS, 1999)
(a) *Chenopodium album* (b) *Portulaca* sp.
(c) *Caprella bersa-pastoris* (d) None of these

125. Vitamin A deficiency causes (ARS, 1999)
(a) Scurvy (b) Beriberi
(c) Night blindness (d) Bleeding of gums

126. Which dose of Alachlor is recommended for weed control in tomato ? *(ARS, 1999)*
(a) 1 to 1.5 Kg a.i/ha (b) 2 to 2.5 Kg. a.i/ha
(c) 3 to 4.0 Kg. a.i/ha (d) None of these

127. The classes of quality seed are *(ARS, 1999)*
(a) Two (b) Three
(c) Four (d) Five

128. The vegetable whose all parts are used as food is *(ARS, 1999)*
(a) Ginger (b) Turmeric
(c) Colocasia (d) Potato

129. Arka Jyoti F_1 hybrid developed by IIHR, Bangalore belongs to *(ARS, 1999)*
(a) Bottle gourd (b) Cucumber
(c) Watermelon (d) Muskmelon

130. The vegetable crop propagated through both seed and vegetative parts is *(ARS, 1999)*
(a) Garlic (b) Tomato
(c) Potato (d) Ginger

131. An individual with one chromosome less than the somatic chromosome number is called as *(ARS, 1999)*
(a) Nullisomic (b) Monosomic
(c) Trisomic (d) None of these

132. Japanese Long Green is a variety of *(ARS, 1999)*
(a) Cabbage (b) Squash
(c) Radish (d) Cucumber

133. Present Dy. Director General (Horticulture) of ICAR is *(ARS, 1999)*
(a) Dr. Peter (b) Dr. H.P. Singh
(c) Dr. Kirti Singh (d) Dr. Ghosh

134. Pusa Meghdoot is the hybrid of *(ARS, 1999)*
(a) Cucumber (b) Bitter gourd
(c) Bottle gourd (d) Pumpkin

135. Pusa Alankar is the hybrid variety of *(ARS, 1999)*
(a) Ashgourd (b) Summer squash
(c) Musk melon (d) Water melon

136. Pant Anupama is a variety of *(ARS, 1999)*
(a) Radish (b) French bean
(c) Carrot (d) Turnip

137. P-8 which is suitable for growing during rainy season in plains belongs to *(ARS, 1999)*
(a) Cauliflower (b) Okra (Bhindi)
(c) Potato (d) Carrot

138. Lines identical in genetic constitution except for one gene are called *(ARS, 1999)*
(a) Pure lines (b) Inbred
(c) Isogenic lines (d) Homozygous lines

139. Which of the breeding methods would be most effective when additive variance is predominant ? *(ARS, 1999)*
(a) Mass selection (b) Pedigree selection
(c) Bulk method
(d) Development of synthetics

140. Which of the following crops is classified as open cross pollinated ? *(ARS, 1999)*
(a) Cucumber (b) Spinach
(c) Fenugreek (d) Chilli

141. Abundance of light and relatively higher temperature increase number of *(ARS, 1999)*
(a) Male flowers (b) Female flowers
(c) Hermaphrodite flowers (d) None of these

142. Little leaf is a disease of *(ARS, 1999)*
(a) Radish (b) Pea
(c) Tomato (d) Brinjal

143. Foundation seed should have genetic purity of *(ARS, 1999)*
(a) 94% (b) 96%
(c) 98% (d) 99.5%

144. Potato was first introduced in India in the year *(ARS, 1999)*
(a) 1575 (b) 1615
(c) 1675 (d) 1715

145. The country is divided from vegetable point of view into *(ARS, 1999)*
(a) 6 Agroclimatic zones (b) 8 Agroclimatic zones
(c) 10 Agroclimatic zones (d) 12 Agroclimatic zones

146. Protandrous type of flowers are present in *(ARS, 1999)*
(a) Cauliflower (b) Onion
(c) Okra (d) Bottle gourd

147. Which of the elements plays indispensable role in the energy metabolism of vegetable plants ? *(ARS, 1999)*
(a) Fe (b) P
(c) N (d) K

148. Which of the bioregulators improves fruit set in chilli ? *(ARS, 1999)*
(a) GA-3 (b) IAA
(c) IBA (d) NAA

149. Most of the potato cultivars in India are *(ARS, 1999)*
   (a) Autotetraploids (b) Allopolyploids
   (c) Diploids (d) Triploids

150. The country having highest area under poly house cultivation is *(ARS, 1999)*
   (a) India (b) China
   (c) Japan (d) America

151. Soilless culture is also known as *(ARS, 1999)*
   (a) NFT culture (b) Nutriculture
   (c) Organic culture (d) None of these

152. The weedicide used for controlling mautha (Cyperus) is *(ARS, 1999)*
   (a) Glyfosate (b) Sulpley
   (c) 2, 4-D (d) None of these

153. M.O.P. contains per cent K_2O *(ARS, 1999)*
   (a) 60 (b) 62
   (c) 70 (d) None of these

154. Vegetable commonly grown under fencing structures are *(ARS, 1999)*
   (a) Cucumber (b) Tomato
   (c) Both (d) Brinjal & Cucumber

155. Punjab Padmini is a variety of okra developed by *(ARS, 1999)*
   (a) C.S.A.U. (b) P.A.U.
   (c) B.H.U. (d) I.I.V.R.

156. Contender is a variety of *(ARS, 1999)*
   (a) French bean (b) Cowpea
   (c) Cluster bean (d) None of these

157. 'Lincon' is the variety of *(ARS, 1999)*
   (a) Pea (b) Bean
   (c) Cow pea (d) Mungbea

158. Angoorlata is the variety of *(ARS, 1999)*
   (a) Grape (b) Tomato
   (c) Hyacinth bean (d) None of these

159. The term "Evergreen Revoluation" has been given by *(ARS, 2003)*
   (a) Dr. A.S. Faroda (b) Dr. M.S. Swaminathan
   (c) Dr. V.L. Chopra (d) Dr. R.S. Paroda

160. Which form of nitrogen is available in urea ? *(ARS, 2003)*
   (a) Ammoniacal (b) Amide
   (c) Nitrate (d) Nitrite

161. Bacterial diseases are controlled by the use of chemicals *(ARS, 2003)*
(a) Antibiotic (b) Viricides
(c) Fungicides (d) Kelthane

162. Application of potash increases *(ARS, 2003)*
(a) Disease resistance in plants
(b) Tolerance to water logging
(c) Frost tolerance in plants (d) None of these

163. How much seed of legume crops should be treated with one packet of Rhizobium culture ? *(ARS, 2003)*
(a) 5 kg (b) 10 kg
(c) 15 kg (d) 20 kg

164. Phosphate containing mineral is *(ARS, 2003)*
(a) Dolomite (b) Aptatite
(c) Marble (d) Pyrites

165. LEISA is related to *(ARS, 2003)*
(a) Natural farming (b) Organic farming
(c) Inorganic farming (d) None of these

166. Which one of the following is a green house gas ? *(ARS, 2003)*
(a) Oxygen (b) Ammonia
(c) Methane (d) None of these

167. 'Tarai' soils are deficient in *(ARS, 2003)*
(a) N (b) P
(c) Zn (d) S

168. In a good quality Farm yard manure C/N ratio should not be more than *(ARS, 2003)*
(a) 40 (b) 30
(c) 20 (d) 10

169. Zero tillage system was first used successfully in 1950 in pasture renovation in *(ARS, 2003)*
(a) Germany (b) Japan
(c) United Kingdom (d) USA

170. How many agroclimatic zones are categorized in India ? *(ARS, 2003)*
(a) 20 (b) 18
(c) 14 (d) 15

171. How many agro-ecological regions are categorized in India ? *(ARS, 2003)*
(a) 120 (b) 21
(c) 14 (d) 15

172. One tonne is equal to *(Ph.D. VGS, 2003)*
(a) 100 kg (b) 907 kg
(c) 1000 kg (d) 2240 kg

173. For calculating value of P, P_2O_5 is multiplied by *(Ph.D. VGS, 2003)*

(a) 3.258 (b) 2.146
(c) 0.437 (d) None of these

174. Which element is considered as energy currency for the plant ? *(Ph.D. VGS, 2003)*

(a) N (b) P
(c) K (d) None of these

175. Among the families of crop plants, the sulphur requirement increases in the order of *(Ph.D. VGS, 2003)*

(a) *Leguminosae>Cruciferae>Graminae*
(b) *Graminae> Cruciferae >Leguminosae*
(c) *Graminae < Leguminosae < Cruciferae*
(d) None of these

176. Maximum nitrification takes place at *(Ph.D. VGS, 2003)*

(a) 10% O_2 (b) 20% O_2
(c) 30% O_2 (d) None of these

177. P content in DAP is *(Ph.D. VGS, 2003)*

(a) 20% (b) 30%
(c) 40% (d) 50%

178. Stem nodulating *Rhizobium* spp. is *(Ph.D. VGS, 2004)*

(a) Sino-rhizobium (b) *AZO-rhizobium*
(c) Both (d) None of these

179. Which one of the following is the permanent method of preservation of vegetables ? *(Ph.D. VGS, 2004)*

(a) Pasteurization (b) Blanching
(c) Canning and bottling (d) Exlusion of moisture

180. Vegetables are canned in *(Ph.D. VGS, 2004)*

(a) Brine (b) Syrup
(c) Distilled water (d) None of these

181. For biological control of tomato fruit borer, the ideal trap crop is *(Ph.D. VGS, 2004)*

(a) Marigold (b) Mustard
(c) Tobacco (d) None of these

182. Exhausting temprature of canned fruits and vegetables is *(Ph.D. VGS, 2004)*

(a) 70°C (b) 80°C
(a) 90°C (a) 100°C

183. The richest source of iron is *(Ph.D. VGS, 2004)*

(a) Spinach (b) Celery
(c) Parsley (d) Peas

184. National Horticulture Board was established in *(Ph.D. VGS, 2004)*
(a) 1976 (b) 1984
(c) 1990 (d) 1995

185. Agriculture and Processed Foods Product Export Development Authority (APEDA) came into existence in *(Ph.D. VGS, 2004)*
(a) 1975 (b) 1980
(c) 1985 (d) 1990

186. Cytoplasmic male sterility is best suited for hybrid seed production in *(Ph.D. VGS, 2004)*
(a) Self pollinated crops (b) Cross pollinated crops
(c) Vegetatively propagated crops
(d) All the above

187. Pure line variety is a *(Ph.D. VGS, 2004)*
(a) Homozygous - Heterogeneous
(b) Homozygous - Homogeneous
(c) Heterogeneous – Heterogeneous
(d) None of these

188. NBPGR is located at *(Ph.D. VGS, 2004)*
(a) Lucknow (b) New Delhi
(c) Kanpur (d) Karnal

189. In which vegetable crop, the first trasgenic was developed ? *(Ph.D. VGS, 2004)*
(a) Cucumber (b) Pumkin
(c) Tomato (d) Potato

190. 'Flavr Savar' is associated with *(Ph.D. VGS, 2004)*
(a) Potato (b) Tomato
(c) Chilli (d) Pomato

191. 'RFLP' is a *(Ph.D. VGS, 2004)*
(a) Genetic marker (b) Molecular marker
(c) Morphological marker (d) Physiological maker

192. The father of genetics was *(Ph.D. VGS, 2004)*
(a) Mendel (b) Kolreuter
(c) Darwin (d) None of these

193. The word 'Mutation' was first used by *(Ph.D. VGS, 2004)*
(a) Muller (b) Hugo de Vries
(c) Wright (d) Morgan

194. The self compatibility is very common in plant kingdom was reported by *(Ph.D. VGS, 2004)*
(a) East (b) Lewis
(c) Mathur (d) None of these

195. Quality seed distributed to farmers for commercial cultivation is *(Ph.D. VGS, 2004)*
(a) Nucleus (b) Breeder
(c) Foundation (d) Certified

196. Recommended concentration of EMS to induce mutation is *(Ph.D. VGS, 2004)*
(a) 0.03-1.5% (b) 0.30-3.0%
(c) 1.00-3.5% (d) All the above

197. The cDNA libraries can be prepared by isolating *(Ph.D. VGS, 2004)*
(a) tRNA (b) mRNA
(c) rDNA (d) DNA

198. The molecular marker which is co-dominant in nature is *(Ph.D. VGS, 2004)*
(a) RAPD (b) SSR and RFLP
(c) VNTR (d) None of these

199. Self splicing introns are found in *(Ph.D. VGS, 2004)*
(a) tRNA (b) rRNA genes
(c) Structural genes (d) In all of these

200. Nullisomic condition is represented by *(Ph.D. VGS, 2004)*
(a) 2n + 1 (b) 2n+1-1
(c) 2n-2 (c) 2n-1

201. The combination of male and female features in body of an individual is known as *(Ph.D. VGS, 2004)*
(a) Sex mosaic (b) Sex Chromatin
(c) Autosomes (d) None of these

202. The eco-friendly insecticide is *(Ph.D. VGS, 2004)*
(a) Malathion (b) Thiodan
(c) Nimbicidine (d) None of these

203. The highest pesticide used in the crop in India is *(Ph.D. VGS, 2004)*
(a) Rice (b) Cotton
(c) Sugarcane (d) Potato

204. The insecticide which can be safely recommended for vegetables is *(Ph.D. VGS, 2004)*
(a) Carbaryl (b) Aldicarb
(c) Endrin (d) Malathion

205. Destructive Insect Pest Act (DIPA) was passed in *(Ph.D. VGS, 2004)*
(a) 1914 (b) 1920
(c) 1937 (d) 1968

206. *Bacillus thuringiensis* was discovered by *(Ph.D. VGS, 2004)*
(a) S. Ishiwata (b) Ber liner
(c) Flancher (d) None of these

207. Which insect has developed resistance to Bt in open field pollination ? *(Ph.D. VGS, 2004)*
(a) Pink boll warm (b) Spotted boll worm
(c) Diamond back moth (d) None of these

208. Juvenile hormone (JH) is a terpene secreted by *(Ph.D. VGS, 2004)*
(a) Corpora cardiaca (b) Corpora allata
(c) Culticle (d) None of these

209. The term "third generation pesticides" is given for *(Ph.D. VGS, 2004)*
(a) Juvenile hormone (b) Culticle
(c) Brain hormone (d) None of these

210. *Bacillus thuringiensis* is available in market as *(Ph.D. VGS, 2004)*
(a) Vertilee (b) Dipel, Delfin
(c) Boverin (d) None of these

211. Central Insecticides laboratory (CIL) is located at *(Ph.D. VGS, 2004)*
(a) Faridabad (b) Lucknow
(c) Kanpur (d) Ludhiana

212. Total number of pesticides restricted for use in India is *(Ph.D. VGS, 2004)*
(a) 186 (b) 24
(c) 7 (d) 4

213. Transgenic crop which has highest area in world is *(Ph.D. VGS, 2004)*
(a) Soybean (b) Cotton
(c) Maize (d) Mustard

214. Barnase, barstar and bar genes are used for developing genetically modifed organisms is *(Ph.D. VGS, 2005)*
(a) Cotton (b) Mustard
(c) Potato (d) Tomato

215. In India Genetic Engineering Approval Committee (GEAC) is responsible for monitoring of GM crops come under ministry of *(Ph.D. VGS, 2005)*
(a) Environment & Forest (b) Agriculture
(c) Information & Broadcasting
(d) None of these

216. Work on GM Mustard is carried out by *(Ph.D. VGS, 2005)*
(a) Monsanto (b) Mahyco
(c) Syngenta (d) Pro Agro

217. The pesticide suitable for controlling the white grub is *(Ph.D. VGS, 2005)*
(a) Phorate (b) Malathion
(c) Chlorpyriphos (d) None of these

218. VAM is *(Ph.D. VGS, 2005)*
(a) Bacteria (b) Fungi
(c) Virus (d) Algae

219. The fungicide which is not systemic in nature is *(Ph.D. VGS, 2005)*
(a) Vitavax (b) Thiram
(c) Benlete (d) Topsin

220. All the plant viruses (except cauliflower mosaic virus) contain *(Ph.D. VGS, 2005)*
(a) DNA (b) RNA
(c) Both (d) None of these

221. Quick and accurate detection of viruses can be done by *(Ph.D. VGS, 2005)*
(a) ELISA (b) IEM
(c) HADAS (d) All the above

222. Sulphur dioxide is toxic to plant at or above concentration of *(Ph.D. VGS, 2005)*
(a) 0.1-0.3 PPm (b) 0.3-0.5 PPm
(c) 3-4 PPm (d) 4-8 PPm

223. Most widely used fungicide for smut fungi is *(Ph.D. VGS, 2005)*
(a) Vitavax (b) Plant vax
(c) Dithane M-45 (d) None of these

224. Race specific resistance is called *(Ph.D. VGS, 2005)*
(a) Vertical resistance (b) Horizontal resistance
(c) Plant resistance (d) All the above

225. When plant shows the partial resistance against all the races of pathogen it is called as *(Ph.D. VGS, 2005)*
(a) Vertical resistance (b) Horizontal resistance
(c) None-host resistance (d) All the above

226. The widely used fungicide for the control of powdry mildew is *(Ph.D. VGS 2005)*
(a) Sulphur dust (b) Dithane M-45
(c) Calaxin (d) All the above

227. Which fungicide is used against downy mildew ? *(Ph.D. VGS, 2005)*
    (a) Karathane (b) Oxathin
    (c) Metalaxyl (d) All the above
228. Calcium is an important constituent of *(Ph.D. VGS, 2005)*
    (a) Protein (b) Cell wall
    (c) Chloroplast (d) None of these
229. Transpiration is measured by *(Ph.D. VGS, 2005)*
    (a) Lysimeter (b) Tensiometer
    (c) Potometer (d) None of these
230. The chemical nature of GA_3 is *(Ph.D. VGS, 2005)*
    (a) Phenolic (b) Purine
    (c) Terpene (d) None of these
231. Which part of root absorbs water and minerals ? *(Ph.D. VGS, 2005)*
    (a) Root cap (b) Root hairs
    (c) Epidermis (d) None of these
232. Growing of plant in soilless nutrient solution is referred as *(Ph.D. VGS, 2005)*
    (a) Hydroponics (b) Aeroponics
    (c) Xeroponics (d) None of these
233. The 'stem crak' of celery, 'heart rot' of beet, 'water core' of turnip are the deficiency symptoms of *(Ph.D. VGS, 2005)*
    (a) Ca (b) B
    (c) Zn (d) All the above
234. The micronutrient essential for the synthesis of 'auxin' is *(Ph.D. VGS, 2005)*
    (a) Zn (b) Cu
    (c) Mn (d) Mo
235. Mycorrhiza increases the availability of *(Ph.D. VGS, 2005)*
    (a) Bo (b) P
    (c) N (d) K
236. VAM (Vesicular arbuscular mycorrhiyzae) is mostly used in *(Ph.D. VGS, 2005)*
    (a) Perennial trees (b) Biennial crops
    (c) Annual crops (d) All the above
237. RNA stands for *(Ph.D. VGS, 2005)*
    (a) Heterogenous nuclear RNA
    (b) Homogeneous nuclear RNA
    (c) Both (c) None of these
238. Recognition site of t-RNA is *(Ph.D. VGS, 2005)*
    (a) Loop I (b) Loop II
    (c) Loop IV (d) Anticodon

239. Chlorophyll is green because it (Ph.D. VGS, 2005)
(a) Reflects green light (b) Transmits green light
(c) Absorbs green light (d) None of these

240. RNA-DNA hybridization to quantity gene expression of mRNA level is called as (Ph.D. VGS, 2005)
(a) Northern blotting (b) Western blotting
(c) Southern blotting (d) None of these

241. Natural inhibitor of IAA oxidases is (Ph.D. VGS, 2005)
(a) ABA (b) Lactic Acid
(c) Coumaric acid (d) None of these

242. In C_4 plant, the first stable product of photosynthesis is (Ph.D. VGS, 2005)
(a) GA (b) PGA
(c) Oxalic Acid (d) Malic acid

243. The present level of CO_2 in atmosphere is (Ph.D. VGS, 2005)
(a) 210-250 ppm (b) 300-320 ppm
(c) 360-370 ppm (d) 420-460 ppm

244. Lysimeter is used in the measurement of (Ph.D. VGS, 2005)
(a) Light (b) Transpiration
(c) Lysine content (d) None of these

245. The transpiration ratio is higher for (Ph.D. VGS, 2005)
(a) C_3 plants (b) C_4 plants
(c) Both (d) None of these

246. The end product of glycolysis is (Ph.D. VGS, 2005)
(a) Glucose (b) Pyrubic acid
(c) Sucrose (d) NADH

247. Glycolysis takes place in (Ph.D. VGS, 2005)
(a) Cytoplasm (b) Chloroplast
(c) Nucleus (d) None of these

248. Kreb cycle produces (Ph.D. VGS, 2005)
(a) 18 ATP (b) 30 ATP
(c) 32 ATP (d) 36 ATP

249. Storage protein in beans is (Ph.D. VGS, 2005)
(a) Insulin (b) Globulins
(c) Rhaseoline (d) None of these

250. Green house gas for global warming is (Ph.D. VGS, 2005)
(a) O_2 (b) CO_2
(c) SO_2 (d) CH_4

251. How many electrons are required for conversion of No_2^- to NH_4^+? (Ph.D. VGS, 2005)
(a) 2 (b) 4
(c) 6 (d) 8

252. Nitrogen fixation is carried out by enzyme *(Ph.D. VGS, 2005)*
(a) Nitrophase (b) Nitrogenase
(c) Both (d) None of these

253. The phosphorus mobilizer is *(Ph.D. VGS, 2005)*
(a) BGA (b) VAM
(c) Clostridium (d) All the above

254. cDNA stands for *(Ph.D. VGS, 2005)*
(a) Copy DNA (b) Cyclic DNA
(c) Complementary DNA (d) None of these

255. Dry weight is commonly obtained by drying the freshly harvested plant material at *(Ph.D. VGS, 2005)*
(a) 50-60°C (b) 60-70°C
(c) 70-80°C (d) 80-100°C

256. The term auxin was first used in 1926 by *(Ph.D. VGS, 2005)*
(a) Jacobs (b) Fristwent
(c) Yabuta (d) None of these

257. IAA is chemically similar to the amino acid *(Ph.D. VGS, 2005)*
(a) Proline (b) Tryptophan
(c) Serine (d) None of these

258. Apical dominance is the result of *(Ph.D. VGS, 2005)*
(a) Auxins (b) Cytokinins
(d) GA_3 (d) None of these

259. Zeatin had first been identified by *(Ph.D. VGS, 2005)*
(a) Skoog (b) Letham
(c) Jhonson (d) None of these

260. The plant hormone volatile in nature is *(Ph.D. VGS, 2005)*
(a) ABA (b) Ethylene
(c) Cytokinins (d) None of these

261. The term vernalization is related to *(Ph.D. VGS, 2005)*
(a) low temperature promotion of early germination
(b) high temperature promotion of early flowering
(c) low temperature promotion of flowering
(d) None of these

262. Off season flowering in plants is positive by giving treatment of *(Ph.D. VGS, 2005)*
(a) Vernalization (b) Photoperiodism
(c) Both (d) None of these

263. Maleic hydrazide (MH) is used to *(Ph.D. VGS, 2005)*
(a) Induce seed germination
(b) Suppress flowering and emergence of suckers
(c) Enhance ripening (d) All of the above

264. The leafy vegetable which is highly successful in summer season is *(Ph.D. VGS, 2005)*
(a) Spinach (b) Indian spinach (palak)
(c) Amaranth (d) Fenugreek

265. Storing of fruits and vegetables in poly bags prolongs the shelf life due to *(UP PCS (Pre), 2006)*
(a) Increase in CO_2 content
(b) Reduction in rate of respiration
(c) Increase in humidity (d) All of the above

266. The group containing C_3 plants is *(UPPCS (Pre), 2006)*
(a) Amranthus, Lettuce (b) Carrot, Amranthus
(c) Tomato, Lettuce (d) Amaranthus, Potato

267. The permanent method of preservation of vegetables is
(a) Pasteurization (b) Blanching
(c) Canning and bottling (d) Exclusion of moisture

268. Which of the following preservatives is used the preparation of tomato ketchup ?
(a) Sodium benzoate (b) Citric acid
(c) Potassium metabisulphate
(d) Sodium metabisulphate

269. Which of the following pairs is not correctly matched ?
(a) Onion—bulb
(b) Garlic—clove
(c) Sweet potato—tuber
(d) Cauliflower—head

270. ....is a phenolic factor present in onion which has antifungal properties.
(a) Quercetin (b) Srinigrin
(c) Catechol (d) Allyl popyldisulphate

271. Which of the vegetables is not transplanted ?
(a) Tomato (b) Radish
(c) Brinjal (d) Cauliflower

272. A stem vegetable is *(ARS, 2003)*
(a) Carrot (b) Knol-khol
(c) Sweet potato (d) Radish

273. Root, leaves, succulent stem and fruits are used as vegetables in *(Ph.D. VGS, 2006)*
(a) Radish (b) Turnip
(c) Knol-khol (d) Celery

274. Agmark was established in
(a) 1927 (b) 1937
(c) 1947 (d) 1949

ANSWERS

1.	(b)	2.	(c)	3.	(c)	4.	(a)	5.	(d)
6.	(a)	7.	(c)	8.	(b)	9.	(a)	10.	(a)
11.	(c)	12.	(d)	13.	(b)	14.	(c)	15.	(a)
16.	(b)	17.	(d)	18.	(b)	19.	(b)	20.	(d)
21.	(b)	22.	(d)	23.	(b)	24.	(d)	25.	(b)
26.	(c)	27.	(b)	28.	(b)	29.	(b)	30.	(c)
31.	(b)	32.	(d)	33.	(d)	34.	(b)	35.	(b)
36.	(a)	37.	(a)	38.	(c)	39.	(c)	40.	(b)
41.	(b)	42.	(c)	43.	(a)	44.	(d)	45.	(b)
46.	(c)	47.	(d)	48.	(b)	49.	(a)	50.	(c)
51.	(a)	52.	(d)	53.	(a)	54.	(c)	55.	(c)
56.	(d)	57.	(b)	58.	(c)	59.	(d)	60.	(b)
61.	(c)	62.	(c)	63.	(b)	64.	(a)	65.	(c)
66.	(d)	67.	(d)	68.	(d)	69.	(c)	70.	(a)
71.	(b)	72.	(c)	73.	(d)	74.	(a)	75.	(a)
76.	(a)	77.	(a)	78.	(a)	79.	(c)	80.	(a)
81.	(d)	82.	(d)	83.	(a)	84.	(a)	85.	(a)
86.	(d)	87.	(c)	88.	(a)	89.	(b)	90.	(a)
91.	(a)	92.	(a)	93.	(b)	94.	(b)	95.	(a)
96.	(c)	97.	(c)	98.	(a)	99.	(c)	100.	(a)
101.	(a)	102.	(a)	103.	(b)	104.	(b)	105.	(c)
106.	(b)	107.	(a)	108.	(c)	109.	(b)	110.	(b)
111.	(c)	112.	(c)	113.	(c)	114.	(a)	115.	(a)
116.	(c)	117.	(d)	118.	(c)	119.	(c)	120.	(a)
121.	(a)	122.	(b)	123.	(b)	124.	(a)	125.	(c)
126.	(b)	127.	(c)	128.	(c)	129.	(c)	130.	(c)
131.	(b)	132.	(d)	133.	(b)	134.	(c)	135.	(b)
136.	(b)	137.	(b)	138.	(c)	139.	(b)	140.	(d)
141.	(b)	142.	(d)	143.	(d)	144.	(b)	145.	(b)
146.	(b)	147.	(c)	148.	(d)	149.	(a)	150.	(c)
151.	(b)	152.	(a)	153.	(a)	154.	(c)	155.	(b)
156.	(a)	157.	(a)	158.	(b)	159.	(b)	160.	(b)
161.	(a)	162.	(a)	163.	(b)	164.	(b)	165.	(b)
166.	(c)	167.	(c)	168.	(d)	169.	(d)	170.	(d)
171.	(b)	172.	(c)	173.	(c)	174.	(b)	175.	(c)
176.	(b)	177.	(a)	178.	(b)	179.	(b)	180.	(a)
181.	(a)	182.	(b)	183.	(a)	184.	(b)	185.	(c)
186.	(b)	187.	(b)	188.	(b)	189.	(d)	190.	(b)
191.	(b)	192.	(a)	193.	(b)	194.	(b)	195.	(d)

196. (a)	197. (b)	198. (b)	199. (b)	200. (c)
201. (a)	202. (c)	203. (b)	204. (d)	205. (a)
206. (a)	207. (c)	208. (b)	209. (a)	210. (b)
211. (a)	212. (c)	213. (a)	214. (b)	215. (a)
216. (d)	217. (a)	218. (b)	219. (b)	220. (b)
221. (d)	222. (b)	223. (a)	224. (a)	225. (b)
226. (c)	227. (c)	228. (b)	229. (c)	230. (c)
231. (b)	232. (a)	233. (b)	234. (a)	235. (b)
236. (a)	237. (a)	238. (d)	239. (c)	240. (a)
241. (c)	242. (c)	243. (c)	244. (b)	245. (a)
246. (b)	247. (a)	248. (b)	249. (c)	250. (b)
251. (c)	252. (b)	253. (b)	254. (c)	255. (c)
256. (b)	257. (b)	258. (a)	259. (b)	260. (b)
261. (c)	262. (b)	263. (b)	264. (b)	265. (d)
266. (c)	267. (d)	268. (a)	269. (d)	270. (c)
271. (b)	272. (b)	273. (a)	274. (b)	

CHAPTER 2

TUBER CROPS

2.1 Potato

- This is an important tuber producing vegetable.
- The highest area and production of potato is in Uttar Pradesh followed by West Bengal and Bihar.
- The highest productivity of potato is in West Bengal followed by Gujarat.
- About 50% of potato produced in the world is utilized as human food.
- Potato is grown in India in almost all the states except Kerala.
- Potato is a dicot plant.
- Potato is a self-pollinated crop but is vegetatively propagated.
- Potato flowers are hermaphrodite (bisexual).
- Kufri Sinduri is a cross between Kufri Red and Kufri Kundan.
- Kufri Badshah is a cross between Kufri Jyoti and Kufri Alankar.
- Kufri Bahar is a cross between Kufri Red and Ginake.
- Kufri Pukhraj is resistant to early blight.
- Kufri Chipsona 1 and Kufri Chipsona 2 are suitable for processing and producing light colour chips and finger fries.
- Maximum area under potato is in alluvial soil followed by hills, black and red soils.
- Generally, the potato crop is grown in India when the maximum day temperature is not above 20°C.
- Potato tubers have a dormancy of nearly 8-10 weeks.
- Seed Plot Technique, essentially consists of raising the healthy seed crop during low aphid period available in northern plains during October-January and cutting the haulms and harvesting the crop before they cross damaging level. This technique was developed by Dr. Pushkarnath.
- Currently, the Central Potato Research Institute (CPRI), Shimla has the sole responsibility of producing breeder's seed in the country.

- CPRI was established in 1949 with headquarters at Patna which was later shifted to Shimla in 1956.
- The All India Coordinated Potato Improvement Project (AICPIP) was started in 1971 with its headquarters at CPRI, Shimla (H.P.).
- The International Potato Center (IPC) is an autonomous scientific institution at Lima in Peru, established in 1971.

2.2 Sweet potato

- *Ipomoea batatas* L; the cultivated sweet potato, belongs to the family *Convolvulaceae*.
- China ranks first in area and production of sweet potato.
- *Batatas* is a hexaploid having 90 chromosomes.
- Anthesis in sweet potato takes place very early in the morning between 4-5 am.
- Sweet potatoes are moderately drought-tolerant.
- Sweet potato is usually vegetatively propagated through vine cuttings.
- Pusa Harit is a heavy yielder and has very late bolting habit as it has low chilling requirement.
- Pusa Bharati has higher vitamin C and β-carotene content.

2.3 Yams

- Dioscorea species commonly known as yams, belongs to the family *Dioscoreaceae* under monocotyledons.
- Yam plants have tuber or rhizome from which the vines and roots emerge annually during the growing season.
- Sree Shilpa is the first hybrid variety identified and released in *D. alata*.
- Yams are propagated vegetatively through tuber.
- A seed tuber weight of 200-250 g is ideal for optimum production in *D. alata*. For *D. esculenta,* a seed tuber yam weight of 100-125 g is optimum.
- In yam, sprouting is delayed by soaking the tubers for 10 hrs in 0.1% solution of maleic hydrazide.

2.4 Artichoke

- The Globe artichoke is a herbaceous perennial in which immature flower heads or buds are used as vegetable.
- Green Globe and Purple Globe are popular cultivars of artichoke.

2.5 Tapioca

- Cassava (*Manihot esculenta* Crantz.), variously called tapioca, manioc, mandioca and yaca, is a major starchy root crop of the tropics.
- Nigeria is the major country growing cassava. Kerala, where the crop was first introduced in India, accounts for 50% of area under cassava.
- The highest production and productivity of cassava is noted in Tamil Nadu.
- Cassava is highly heterozygous.
- Cassava is a monoecious plant producing both male and female flowers on the same inflorescence and is cross pollinated.
- The main constituent of the cassava tuber is starch.
- The bitter constituent of cassava is cyanogenic glucoside (HCN).
- The yellow colour of the cassava flesh is due to the presence of carotene, the precursor of vitamin A.

2.6 Elephant foot yam

- Elephant's foot yam (*Amorphophallus companulatus*) or '*suran*' or '*jimmikand*', is gaining popularity because of its yield potential and culinary properties.
- Gajendra and Santragachi are the non-acrid, high-yielding varieties of elephant foot yam.
- High-yielding, non-acrid varieties should be propagated vegetatively by using either cut tubers or small tubers.
- For commercial cultivation, whole or cut tubers 500-1,000 g are used for planting.
- Whole tubers should be preferred over cut tubers to minimize rotting.
- Water stagnation during crop growth results in tubers that are very hard to cook. Hence, water stagnation should not be allowed at any stage of crop growth.

2.7 Colocasia

- Colocasia (*Colocasia esculenta*) or taro is the most important and one of the oldest crops.
- The eddoe type is commonly called *arvi* and dasheen as *banda.* Eddoe type is most prevalent as vegetable.
- Colocasia tubers are rich in starch.
- Colocasia leaves and petioles are also used as vegetables.
- Satamukhi, Sree Rashmi and Sree Pallavi are improved varieties of colocasia.

- Colocasia is propagated vegetatively, mostly by small cormels weighing 20-25 g.
- Yam bean (*Pachyrrhizus erosus*) is popularly known as potato bean and mishrikand.
- Mexican type cultivars have larger tubers (500-700 g), whereas local type cultivars have smaller tubers (200-300 g).
- Rajendra Mishrikand 1, an improved variety, is very popular in north Bihar.
- Yam bean is propagated usually through seeds.
- Cracking of tubers is the main problem due to unavailability of proper soil moisture. Delay in harvesting also causes cracking in tubers.
- Availability of proper soil moisture during 45-90 days checks cracking. Application of potassium reduces cracking of tubers.

The objective type questions on tuber crops are given below :

1. An early maturing variety of potato which does not degenerate rapidly is
   (a) Kufri Chandramukhi (b) Kufri Sinduri
   (c) Kufri Alankar (d) None of the above
2. Which one of the following potato varieties has very high Bulking rate ?
   (a) Kufri Naveen (b) Kufri Alankar
   (c) Kufri Chandramukhi (d) Kufri Sinduri
3. Early and late blights of potato are caused by
   (a) Bacteria (b) Fungi
   (c) Virus (d) None of the above
4. Earthing of potato crop is done after
   (a) 20 days of sowing (b) 40 days of sowing
   (c) 50 days of sowing (d) 60 days of sowing
5. Exceptionally large potato tubers have more chances of
   (a) Brown-heart (b) Hollow-heart
   (c) Black-heart (d) All the above
6. For low sugar content, potato tubers should be stored at the temperature of
   (a) 6°C (b) 10°C
   (c) 15°C (d) 20°C
7. Potato tubers should be kept for long storage at the temperature of
   (a) 10-15°C (b) 15-20°C
   (c) 20-25°C (d) Above 25°C

8. Golden nematode is a serious pest of
   (a) Cabbage (b) Potato
   (c) Tomato (d) All the above
9. Requirement of irrigation water for optimum potato production is
   (a) 500 mm (b) 600 mm
   (c) 700 mm (d) 800 mm
10. Staple vegetable in our diet is
   (a) Tomato (b) Potato
   (c) Brinjal (d) Okra
11. In potato, hollow heart is due to
   (a) Excess water and high temperature
   (b) Excess water and fertilization
   (c) Excess water and low temperature
   (d) Excess water and Nitrogen
12. Seed rate per hectare of potato is
   (a) 5 quintals (b) 10 quintals
   (c) 15 quintals (d) 20 quintals
13. Kufri Badshah a variety of potato is resistant to
   (a) Late blight (b) Early blight
   (c) Wart disease (d) Viral disease
14. Kufri Sinduri is cross between
   (a) Kufri Red and Phulwa (b) Kufri Red and Kundan
   (c) Kufri Red and Sutlej (d) Kufri Red and Alankar
15. Kufri Badshah is a cross between
   (a) Kufri Red and Phulwa (b) Kufri Red and Kundan
   (c) Kufri Jyoti and Kufri Alankar
   (d) None of the above
16. Kufri Swarna is highly resistant to
   (a) Late blight (b) Cyst nematode
   (c) Both (d) None of the above
17. Planting time of three crops of potato grown in the Nilgiri hills is
   (a) September, February and July
   (b) April, August and January
   (c) October, December and March
   (d) None of the above
18. Planting space of potato crop is
   (a) 75 x 10 cm (b) 60 x 10 cm
   (c) 60 x 15 cm (d) None of the above

19. Potato is a
(a) Long-day plant (b) Short-day plant
(c) Day-neutral plant (d) None of the above

20. Potato is native of
(a) Europe (b) Africa
(c) South America (d) None of these

21. Potato is a rich source of vitamin
(a) A (b) B
(c) C (d) D

22. Potato is mainly propagated by
(a) Tubers (b) Corm
(c) Seeds (d) Cuttings

23. Potato tuber production is maximum at the temperature of
(a) 15°C (b) 20°C
(c) 25°C (d) 30°C

24. Day temperature requirement for potato production is
(a) 25°C (b) > 25°C
(c) < 40°C (d) < 35°C

25. Night temperature requirement for potato tuber production is
(a) 20°C (b) < 20°C
(c) 20-25°C (d) None of these

26. Potato variety resistant to wart disease is
(a) Kufri Jyoti (b) Kufri Muthu
(c) Kufri Sherpa (d) All the above

27. The ideal soil for potato is one which has pH range of
(a) 4.5-5.0 (b) 5-6.5
(c) 6.5-7.5 (d) None of the above

28. The usual weight of seed piece of potato is
(a) 30-40 g (b) 40-50 g
(c) 50-60 g (d) None of the above

29. The young potato plant best grown at the temperature of
(a) 15°C (b) 20°C
(c) 24°C (d) 28°C

30. Tuber production stops totally at the temperature of
(a) 25°C (b) 30°C
(c) 35°C (d) None of the above

31. Viral disease of potato is spread by
    (a) Jassids (b) White flies
    (c) Aphids (d) None of the above
32. Where do we have to contact to procure fresh seeds of potato ?
    (a) IARI, New Delhi (b) CPRI, Shimla
    (c) Darjeeling (d) Lucknow
33. The seed rate of TPS which is recommended for planting one hectare area is
    (a) 150 g (b) 250 g
    (c) 350 g (d) 400 g
34. A pre-harvest spray with MH on potato controls
    (a) Dormancy (b) Sprouting
    (c) Both (d) None of the above
35. Dormancy of potato tubers can be broken by treating with
    (a) Maleic hydroxide (b) Acetic acid
    (c) Thiourea (d) None of the above
36. Hill potatoes are nearly free from disease
    (a) Buckeye rot (b) Charcol rot
    (c) Both (d) None of the above
37. In Bihar, Himachal Pradesh and Nilgiri hills of India, potato is grown as a
    (a) Rain fed crop (b) Semi rain fed crop
    (c) Both (d) None of the above
38. In Nilgiri hills in Tamil Nadu, potato is rotated with
    (a) Chikpea (b) Pea
    (c) Finger millet (d) None of the above
39. Potato in Nilgiri hills of Tamil Nadu was introduced in
    (a) 1810 (b) 1820
    (c) 1822 (d) 1830
40. Main crop of potato in hills of Himalayan region is planted in
    (a) January-February (b) March-April
    (c) May-June (d) None of the above
41. Potato crop removes large amount of potassium followed by
    (a) N and P (b) N and Fe
    (c) N and K (d) None
42. Potato requires ------ conditions for its tuber development.
    (a) Long-day (b) Short-day
    (c) Day-neutral (d) None of the above
43. Potato stored at –1 to 0°C suffers from internal break-down known as
    (a) Black-heart (b) Brown-heart
    (c) Hollow-heart (d) None of the above

44. Potato was introduced in Europe by the early Spanish exporters during
    (a) 16th Century (b) 17th Century
    (c) 18th Century (d) None of the above
45. Potato grown in alkaline soils is attacked by
    (a) Viral disease (b) Scab disease
    (c) Wart disease (d) None of the above
46. The potato variety Kufri Kuber has been developed at Potato Breeding Station (now CPRI), Shimla in
    (a) 1940 (b) 1941
    (c) 1942 (d) 1943
47. The optimum size of potato seed piece is 40-50 g with diameter of
    (a) 20-30 mm (b) 30-40 mm
    (c) 40-50 mm (d) None of the above
48. The tubers of potato are borne underground at
    (a) Stolen ends (b) Stolen upper side
    (c) Stolen middle side (d) None of the above
49. The young plants of potato grow best at the temperature of 24°C and later growth is favoured at the temperature of
    (a) 10°C (b) 15°C
    (c) 18°C (d) None of these
50. In the world scenario, more than 80% of the area under sweet potato cultivation and production lies in
    (a) Africa (b) India
    (c) Asia (d) None of these
51. Potato varieties Kufri Alankar, Kufri Jawahar, Kufri Ashoka and Kufri Chandramukhi are
    (a) Early maturing (b) Mid maturing
    (c) Late maturing (d) None of these
52. Potato varieties Kufri Deva and Kufri Megha are
    (a) Early maturing (b) Mid maturing
    (c) Late maturing (d) None of these
53. Introduced variety of potato is
    (a) Uptodate (b) Kufri Chipsona
    (c) Kufri Sutlej (d) None of these
54. Kufri Red and Kufri Safed varieties of potato were developed through
    (a) Pure line (b) Mass selection
    (c) Clonal selection (d) None of these

55. Kufri Chipsona-2 is tolerant to
(a) Frost (b) Late blight
(c) Both (d) None of these

56. Potato variety immune to wart disease and susceptible to cracking is
(a) Kufri Sherpa (b) Kufri Jyoti
(c) Both (d) None of these

57. Kufri Swarna is resistant to
(a) Late blight (b) Cyst nematode
(c) Both (d) None of these

58. Multiple disease resistant variety of potato is
(a) Kufri Badshah (b) Kufri Jyoti
(c) Kufri Deva (d) None of these

59. Potato varieties suitable for processing are
(a) Kufri Chipsona-1 (b) Kufri Chipsona-2
(c) Both (d) None of these

60. Potato varieties Kufri Pukhraj and Kufri Chandramukhi are resistant to
(a) Early blight (b) Late blight
(c) Cyst nematode (d) None of these

61. Frost resistant variety of potato is
(a) Kufri Lavkar (b) Kufri Sheetman
(c) Kufri Ashoka (d) None of these

62. CPRI was established in
(a) 1949 (b) 1959
(c) 1939 (d) None of these

63. Protein content in potato is
(a) 1.6% (b) 1.9%
(c) 2.6% (d) None of these

64. Carbohydrate content in potato (per 100 g) is
(a) 23.6 g (b) 22.6 g
(c) 21.6 g (d) None of these

65. Solanin content per 100 g of potato is
(a) 5 mg (b) 7 mg
(c) 9 mg (d) None of these

66. Potato unit for consumption solanin content in per 100 g is
(a) More than 200 mg (b) Less than 200 mg
(c) Equal to 200 mg (d) None of these

67. Wart disease is confined to the Darjeeling hills of
(a) Bihar (b) West Bengal
(c) Orissa (d) None of these

68. Treatment with 1% thiourea + 1 ppm GA_3 for one hour breaks the dormancy of
    (a) Sweet potato (b) Potato
    (c) Yam (d) None of these

69. India's rank in the world is 4th in area and 3rd in production of
    (a) Potato (b) Sweet potato
    (c) Tomato (d) None of these

70. Dehaulming in potato is essential for disease free
    (a) Seed production (b) Tuber production
    (c) Both (d) None of these

71. Seed Plot Technique in potato was developed by
    (a) Dr. G. Kalloo (b) Dr. Pushkarnath
    (c) Dr. S. Ramanujam (d) None of these

72. The area under potato cultivation in plains is
    (a) 85% (b) 95%
    (c) 90% (d) 75%

73. Duration of potato tuber dormancy is
    (a) 6-7 weeks (b) 8-10 weeks
    (c) 11-12 weeks (d) 14-15 weeks

74. The area under potato cultivation in hills is
    (a) 5% (b) 7%
    (c) 9% (d) 11%

75. The country I^{st} in area and production of potato in the world is
    (a) India (b) China
    (c) USA (d) Pakistan

76. Potato is grown in all states of India except
    (a) Kerala (b) Tamil Nadu
    (c) Andhra Pradesh (d) Karnataka

77. Potato crop is a
    (a) Self pollinated (b) Cross pollinated
    (c) Often cross pollinated (d) None of these

78. The country which ranks first in area and production of sweet potato is
    (a) India (b) China
    (c) Africa (d) None of these

79. *Ipomea batatas* L. is a hexaploid having chromosome number
    (a) 70 (b) 80
    (c) 90 (d) None of these

80. Anthesis in sweet potato takes place very early in the morning between
(a) 3-4 a.m. (b) 4-5 a.m.
(c) 5-6 a.m. (d) 7-8 a.m.

81. Sweet potato is often treated as
(a) Annual crop (b) Perennial crop
(c) Both (d) None of these

82. Sweet potato vine is treated as
(a) Annual vine (b) Perennial vine
(c) Both (d) None of these

83. The flowers of sweet potato are
(a) Cymose/Solitary (b) Racemose
(c) Spadix (d) None of these

84. Sweet potato is propagated through
(a) Root cuttings (b) Vine cuttings
(c) Tuber cuttings (d) None of these

85. Each cutting of sweet potato should have at least
(a) 2 Nodes (b) 3 Nodes
(c) 4 Nodes (d) None of these

86. Number of sweet potato cuttings required to plant in one hectare of area is
(a) 20,000-30,000 (b) 30,000-40,000
(c) 40,000-50,000 (d) None of these

87. How many tubers will be needed to grow yam in one hectare area ?
(a) 13,300 (b) 14,000
(c) 15,000 (d) 16,300

88. Elephant foot corm at the end of third year weighs
(a) 2-3 kg (b) 3-4 kg
(c) 5-6 kg (d) 6.8-9.0 kg

89. Elephant foot yam is rich in vitamins
(a) A (b) B
(c) A and B (d) None of these

90. In central India, sweet potato vine cuttings are planted during
(a) May-June (b) June-July
(c) August-September (d) October-November

91. Flower colour of sweet potato is
(a) White (b) Yellow
(c) White to purple (d) None of these

92. The tuber crop which is drought-tolerant is
(a) Sweet potato (b) Potato
(c) Both (d) None of these

93. Low chilling requiring variety of sweet potato is
(a) Pusa Bharati (b) Pusa Harit
(c) Both (d) None of these

94. Sweet potato variety containing higher amount of vitamin C and β-carotene is
(a) Pusa Harit (b) Pusa Bharati
(c) Both (d) None of these

95. *Dioscorea* species are commonly known as
(a) Yam (b) Bean
(c) Both (d) None of these

96. *Dioscorea alata* is known as
(a) Water yam (b) Greater yam
(c) Both (d) None of these

97. *Dioscorea esculenta* is known as
(a) Water yam (b) Greater yam
(c) Lesser or Chinese yam (d) None of these

98. *Dioscorea rotundata* is known as
(a) Lesser yam (b) White yam
(c) Red yam (d) None of these

99. The crop contains diosgenin is
(a) Sweet potato (b) Potato
(c) yam (d) None of these

100. First hybrid variety identified and released in *Dioscorea alata* is
(a) Sree Shilpa (b) Sree Harsha
(c) Both (d) None of these

101. Yams are propagated vegetatively through
(a) Vine cuttings (b) Tubers
(c) Both (d) None of these

102. For economic production yams are propagated through
(a) Vine (b) Tubers
(c) Both (d) None of these

103. A seed tuber weight of *Dioscorea alata* is
(a) 100-150 g (b) 150-200 g
(c) 200-250 g (d) None of these

104. A seed tuber weight of *Dioscorea esculenta* is
(a) 100-125 g (b) 125-150 g
(c) 150-200 g (d) None of these

105. Yam corms generally remain dormant in storage for
(a) 1-1.5 Month (b) 1.5-2.0 Months
(c) 2-2.5 Months (d) None of these

106. Dormancy in yams can be broken by quick dipping in 4-8% solution of
(a) Ethylene chlorohydrin (b) Ethylene
(c) Thiurea (d) None of these

107. In yam, sprouting is delayed by soaking the corms for 8-10 hrs in 0.1% solution of
(a) Maleic hydrazide (b) Sulphur hydraxide
(c) Urea (d) None of these

108. The globe artichoke (*Cynara scolymus*) is a
(a) Herbaceous plant (b) Perennial plant
(c) Both (d) None of these

109. Edible part of globe artichoke is
(a) Immature flower (b) Heads
(c) Buds (d) All the above

110. Green globe and purple globe are popular cultivars of
(a) Yam (b) Artichoke
(c) Cassava (d) None of these

111. Globe artichoke is propagated by
(a) Seed (b) Suckers
(c) Off-shoots (d) All the above

112. In artichoke the thick receptacle known as 'heart' is used for
(a) Eating (b) Canning
(c) Propagation (d) None of these

113. Cassava is also known as
(a) Tapioca (b) Manioc
(c) Mandioca (d) Yaca
(e) All the above

114. The major starch root crop of the tropics is
(a) Potato (b) Sweet potato
(c) Cassava (d) None of these

115. The major cassava growing country is
(a) China (b) India
(c) Nigeria (d) None of these

116. In India the major cassava growing state is
(a) Andhra Pradesh (b) Madhya Pradesh
(c) Assam (d) Kerala

117. The highest production and productivity of cassava is noted from
(a) Assam (b) Kerala
(c) Tamil Nadu (d) None of these

118. Native of cassava is
(a) China (b) India
(c) Brazil (d) None of these

119. Cassava belongs to the family
(a) *Malvaceae* (b) *Euphorbiaceae*
(c) *Cheinopodiaceae* (d) None of these

120. Cassava is a
(a) Self pollinated (b) Cross pollinated
(c) Often cross pollinated (d) None of these

121. Cassava is
(a) Monoecious (b) Dioecious
(c) Both (d) None of these

122. The main constituent of the cassava tuber is
(a) Fat (b) Starch
(c) Vitamins (d) None of these

123. The bitter constituent of cassava is
(a) Cyanogenic glucoside (b) Sulphur
(c) Oxalic acid (d) None of these

124. Yellow colour of cassava flesh is due to
(a) Carotene (b) Vitamin B_{12}
(c) Glucoside (d) None of these

125. Nidhi is a variety of
(a) Artichoke (b) Yam
(c) Cassava (d) Coleus

126. Cassava is propagated through
(a) Seed (b) Tuber
(c) Cuttings (d) None of these

127. Coleus (*Solenostemon rotundifolius* Poir.) is commonly known as
(a) Indian potato (b) Chinese potato
(c) African potato (d) None of these

128. Edible part of coleus which can be used as vegetable is
(a) Fruit (b) Tuber
(c) Seed (d) None of these

129. The genus Coleus belongs to the family
(a) *Labialae* (b) *Araceae*
(c) *Cucurbitaceae* (d) None of these

130. Origin of Coleus is
(a) Africa (b) India
(c) China (d) None of these

131. Sree dhara is a variety of coleus released in
(a) Tamil Nadu (b) Kerala
(c) Assam (d) None of these

132. Coleus is propagated through
(a) Seed (b) Vine cuttings
(c) Tuber (d) None of these

133. Primary product of giant taro is
(a) Aerial (b) Stem
(c) Tuber (d) Fleshy aerial stem

134. The giant taro originated in
(a) Africa (b) Sri Lanka
(c) Asia (d) None of these

135. In Assam and West Bengal giant taro is cultivated as
(a) Seed crop (b) Tuber crop
(c) Food crop (d) None of these

136. Propagation of giant taro is by
(a) Suckers (b) Shoot tips
(c) Cormels (d) All the above

137. *Amorphophallus companulatus* is also known as
(a) Elephants foot (b) Suran
(c) Jimmikand (d) All the above

138. Non-acrid and high-yielding variety of suran is
(a) Gajendra (b) Santragachi
(c) Both (d) None of these

139. Suran is propagated by
(a) Cut tubers (b) Small tubers
(c) Both (d) None of these

140. For commercial cultivation of suran, mean weight of whole or cut tubers used for planting is
(a) 100-200 g (b) 200-300 g
(c) 300-500 g (d) 500-1000 g

141. The dormancy in suran can be broken by
(a) Thiourea (0.1%) (b) GA_3 (0.1%)
(c) Ethrel (0.1%) (d) All the above

142. *Colocasia esculenta* is the botanical name of
(a) Yam (b) Yam bean
(c) Taro (d) None of these

143. The country which ranks first in area and production of Colocasia is
(a) Africa (b) Asia
(c) India (d) None of these

144. Taro tubers are rich in
(a) Fatty acid (b) Starch
(c) carbohydrate (d) None of these

145. Colocasia tubers contain
(a) Protein (b) Minerals
(c) Both (d) None of these

146. Satamukhi, Sree Rashmi and Sree Pallavi are improved varieties of
(a) Yam (b) Cassava
(c) Taro (d) None of these

147. The ideal weight of Colocsia corm for transplanting is
(a) 10-15 g (b) 15-20 g
(c) 20-25 g (d) None of these

148. Seed rate of taro is
(a) 5 q/ha (b) 10 q/ha
(c) 15 q/ha (d) 20 q/ha

149. *Pachyrrhizus erosus* is known as
(a) Yam bean (b) Potato bean
(c) Mishrikand (d) All the above

150. Edible part of yam bean is
(a) Stem (b) Tuber
(c) Leaves (d) None of these

151. Improved variety of yam bean popular in north Bihar is
(a) Shree Kand (b) Rajendra Mishrikand
(c) Both (d) None of these

152. Yam bean is propagated through
(a) Seed (b) Vine
(c) Tuber (d) None of these

153. Seeds of yam bean are toxic due to the substance
(a) Rotenone (b) β-carotene
(c) Sulphur (d) None of these

154. Cracking of tubers in yam bean is a major problem due to
(a) Unavailability of proper soil moisture
(b) Delay in harvesting
(c) Both (d) None of these

155. Garden beet belongs to the family *(ARS, 2003)*
(a) *Cruciferae* (b) *Chenopodiaceae*
(c) *Umbelliferae* (d) None of these

156. Potato variety which is not resistant to wart disease is *(ARS, 2003)*
(a) Kufri Sharpa (b) Kufri Jyoti
(c) Kufri Muthu (d) Kufri Sinduri

157. For long storage, potato tubers should be kept at the temperature of *(ARS, 2003)*
(a) 5-10°C (b) 10-15°C
(c) 15-20°C (d) None of these

158. Vine cuttings of sweet potato are planted at a spacing of *(ARS, 2003)*
(a) 60 x 30 cm (b) 100 x 60 cm
(c) 150 x 60 cm (d) None of these

159. Jerusalem artichoke is native of *(ARS, 2003)*
(a) North America (b) Mediterranean region
(c) Europe (d) None of these

160. Haulms in potato seed crop are cut during *(ARS, 2003)*
(a) End of December or first week of January
(b) Second fortnight of January
(c) First fortnight of February
(d) None of the above

161. The native place of garden beet is *(ARS, 2003)*
(a) Asia (b) America
(c) Europe (d) None of these

162. Kufri Badshah variety of potato is resistant to *(ARS, 1992)*
(a) Early blight (b) Late blight
(c) Viral disease (d) Wart disease

163. Dormancy of potato tubers remains for a period of *(ARS, 1992)*
(a) 1-2 months (b) 2-3 months
(c) 2-5 months (d) 5-6 months

164. The colour of Pusa Sunehri variety of sweet potato is *(Ph.D. VGS, 2006)*
(a) White (b) Yellow
(c) Light orange (d) Black

165. Sweet potato originated in *(Ph.D. VGS, 2006)*
(a) Africa (b) Tropical America
(c) India (d) Nepal

166. Potato tuber is a modified form of *(UP PCS (Pre), 1995)*
(a) Root (b) Stem
(c) Leaf (d) Flower

167. Maximum calorie per 100 gram is obtained from *(UP PCS (Pre), 1996)*
(a) Carrot (b) Radish
(c) Sweet potato (d) Potato

168. The most suitable late blight resistant variety of potato for hills is *(Ph.D. VGS, 2006)*
   (a) Kufri Alankar (b) Kufri Kundan
   (c) Kufri Jyoti (d) Kufri Chandramukhi

169. The edible portion of sweet potato is *(UP PCS (Pre), 1997)*
   (a) Modified root (b) Leaf
   (c) Modified stem (d) Fruit

170. Potato is propagated through *(Ph.D. VGS, 2006)*
   (a) Tubers (b) TPS
   (c) Micro tubers (d) All of these

171. Suitable temperature for tuberization in potato is *(UP PCS (Pre), 1997)*
   (a) 12-14°C (b) 22-24°C
   (c) 18-20°C (d) 28-30°C

172. Chautancy, Pusa Kesar and Coreless are varieties of *(Ph.D. VGS, 2006)*
   (a) Radish (b) Carrot
   (c) Potato (d) Tomato

173. The bitterness in Colocasia corms is due to the presence of *(UP PCS (Pre), 1998, JRF, 1999)*
   (a) Calcium (b) Calcium oxalate
   (c) Capsaicin (d) Sodium chloride

174. The chromosome number of potato is *(UP PCS (Pre), 1999)*
   (a) 2n = 16 (b) 2n = 32
   (c) 2n = 48 (d) 2n = 20

175. Colocasia, tapioca and elephant foot yam are *(Ph.D. VGS, 2006)*
   (a) Bulb crops (b) Root crops
   (c) Leafy vegetables (d) Cole crops

176. The purpose of dehaulming in potato seed crop is *(UP PCS (Pre), 2003, IAS (Pre), 2003)*
   (a) Better quality of seed tuber
   (b) Higher yield
   (c) Higher protein (d) Higher starch

177. Tuber crop rich in protein is *(UP PCS (Pre), 2003)*
   (a) Potato (b) Sweet potato
   (c) Colocasia (d) Yam

178. Potato variety resistant to late blight disease is *(UP PCS (Pre), 2005)*
   (a) Kufri Chandramukhi (b) Kufri Sinduri
   (c) Kufri Bahar (d) Kufri Badshah

179. Low temperature during storage in potato causes *(ARS, 1999)*
(a) Pithyness (b) Sweetness
(c) Chilling injury (d) None of these

180. Jimmikand originated in *(ARS, 1999)*
(a) Malaysia (b) India
(c) Africa (d) None of these

181. Sex expression in yam is *(ARS, 1999)*
(a) Dioecious (b) Monoecious
(c) Both (d) None of these

182. Edible portion of rhubarb is *(ARS, 1999)*
(a) Leaf (b) Petiole
(c) Root (d) None of these

183. The chromosome number (2n) of tapioca is *(ARS, 1999)*
(a) 18 (b) 24
(c) 32 (d) 36

184. Kufri Jyoti is a variety of *(ARS, 1999)*
(a) Cauliflower (b) Cabbage
(c) Potato (d) Pea

185. The first GM potato developed at CPRI for increasing protein content in tubers consists of genes from *(Ph.D. VGS, 2003)*
(a) Chickpea (b) Pigeonpea
(c) Field pea (d) Grain amaranth

186. The main crop of potato is planted from *(Ph.D. VGS, 2003)*
(a) 25th September to 10th October
(b) 15th October to 25th October
(c) 25th October to 25th November
(d) 25th November to 25th December

187. The TPS variety of Potato is *(Ph.D. VGS, 2003)*
(a) JH 222 (b) HPS 1/113
(c) PJ 376 (d) None of these

188. The pest attacking the potato tuber both in field and storage is *(Ph.D. VGS, 2004)*
(a) Potato aphid (b) Lady bird beetle
(c) Potato tuber moth (d) Potato black cut warm

189. Early blight of potato produces *(Ph.D. VGS, 2005)*
(a) Conidia (b) Telia
(c) Jredia (d) None of these

190. For the effective control of early blight of potato, most suitable fungicide is *(Ph.D. VGS, 2005)*
(a) Zineb (b) Dithane M-45
(c) Blitox (d) None of these

191. Early blight of potato is *(Ph.D. VGS, 2005)*
(a) Air borne (b) Soil borne
(c) Both (d) None of these

192. Potato spindle tuber disease is transmitted *(Ph.D. VGS, 2005)*
(a) Biologically (b) Mechanically
(c) Fungi (d) All the above

193. Black heart of potato is the result of *(Ph.D. VGS, 2005)*
(a) High temperature
(b) Poor ventilation in the store
(c) High temperature of soil during growth and tuber maturation
(d) All the above

194. Leaves and petioles are edible in *(Ph.D. VGS, 2005)*
(a) Sweet potato (b) Colocasia
(c) Elephant foot yam (d) Potato

195. Scab of Irish potatoes, Pox of sweet potatoes are favored by *(Ph.D. VGS, 2003)*
(a) Acidic conditions
(b) Highly alkaline conditions
(c) Neutral to alkaline conditions
(d) None of these

196. The length of sweet potato cutting should normally be..... with 4.5 sound (healthy) buds *(ARS, 2003)*
(a) 40 cm (b) 30 cm
(c) 20 cm (d) 10 cm

197. The most important potato growing state is *(ARS, 2003)*
(a) Karnataka (b) Uttar Pradesh
(c) Andhra Pradesh (d) None of these

ANSWERS

1. (a)	2. (b)	3. (b)	4. (b)	5. (b)
6. (b)	7. (b)	8. (b)	9. (c)	10. (b)
11. (d)	12. (c)	13. (a)	14. (b)	15. (c)
16. (c)	17. (b)	18. (c)	19. (a)	20. (c)
21. (c)	22. (a)	23. (b)	24. (d)	25. (b)
26. (d)	27. (b)	28. (b)	29. (c)	30. (b)
31. (c)	32. (b)	33. (a)	34. (b)	35. (c)
36. (b)	37. (a)	38. (c)	39. (c)	40. (b)
41. (a)	42. (b)	43. (a)	44. (a)	45. (b)
46. (d)	47. (c)	48. (a)	49. (c)	50. (c)

51. (a)	52. (c)	53. (a)	54. (c)	55. (c)
56. (c)	57. (c)	58. (a)	59. (c)	60. (a)
61. (b)	62. (a)	63. (a)	64. (b)	65. (a)
66. (a)	67. (b)	68. (b)	69. (a)	70. (a)
71. (b)	72. (a)	73. (b)	74. (a)	75. (b)
76. (a)	77. (a)	78. (b)	79. (c)	80. (b)
81. (a)	82. (b)	83. (a)	84. (b)	85. (c)
86. (c)	87. (a)	88. (d)	89. (c)	90. (d)
91. (c)	92. (a)	93. (b)	94. (b)	95. (a)
96. (c)	97. (c)	98. (b)	99. (c)	100. (a)
101. (c)	102. (b)	103. (c)	104. (a)	105. (c)
106. (a)	107. (a)	108. (c)	109. (d)	110. (b)
111. (d)	112. (b)	113. (e)	114. (c)	115. (c)
116. (d)	117. (c)	118. (c)	119. (b)	120. (b)
121. (a)	122. (b)	123. (a)	124. (a)	125. (c)
126. (c)	127. (b)	128. (b)	129. (a)	130. (a)
131. (b)	132. (b)	133. (d)	134. (b)	135. (c)
136. (d)	137. (d)	138. (c)	139. (c)	140. (d)
141. (d)	142. (c)	143. (a)	144. (b)	145. (c)
146. (c)	147. (c)	148. (b)	149. (d)	150. (c)
151. (b)	152. (a)	153. (a)	154. (c)	155. (b)
156. (d)	157. (b)	158. (a)	159. (a)	160. (a)
161. (c)	162. (b)	163. (b)	164. (c)	165. (b)
166. (b)	167. (c)	168. (c)	169. (a)	170. (d)
171. (c)	172. (b)	173. (b)	174. (c)	175. (b)
176. (a)	177. (c)	178. (d)	179. (c)	180. (b)
181. (a)	182. (b)	183. (d)	184. (c)	185. (d)
186. (b)	187. (b)	188. (c)	189. (a)	190. (b)
191. (b)	192. (b)	193. (d)	194. (b)	195. (c)
196. (b)	197. (b)			

CHAPTER 3

LEAFY VEGETABLES

This is very important group of vegetables which are rich in vitamins and minerals.

3.1 Spinach

- Spinach or *Palak* beet is the most common leafy vegetable of tropical and sub-tropical regions.
- Palak leaves contain low oxalic acid.
- Leaves of cultivar Punjab Green are succulent and free from sourness.
- The cultivars Pusa Jyoti was developed by polyploidization by using All Green as a source material.
- Pusa Harit was developed by hybridization between sugar beet and local palak at IARI Regional Station, Katrain (H.P.).
- Jobner Green was developed by selection from a spontaneous mutation in a local collection, Sl No.5.
- Pusa Bharti is the latest variety developed at, IARI, New Delhi.
- Traditionally, spinach or *vilayati palak* is classified as pot herb. It is green ones that are normally consumed as cooked vegetable.
- Spinach is exceptionally grown in hilly areas and not in plains of India.
- Virginia Savoy is a prickly seeded cultivar of spinach.
- Spinach is a long-day crop.

3.2 Amaranth

- Amaranth is primarily used as pot herb.
- Amaranth is the most common leafy vegetable grown during summer and rainy season in India.
- Among leafy types, *Amaranthus tricolour* L. is the main cultivated species in India.
- Amaranth is a cross pollinated (Anemophilous) crop.
- Pusa Kiran and Pusa Kirti are the varieties of amaranth.

The objective type questions are given below :

1. Amaranth is grown in India during
   (a) Summer season (b) Rainy season
   (c) Both (d) None of these
2. Amaranth is a
   (a) Self pollinated (b) Cross pollinated
   (c) Both (d) None of these
3. *Amaranthus tricolour* L. is cultivated species in
   (a) India (b) China
   (c) South Africa (d) None of these
4. Amaranth is a
   (a) Anemophilous (b) Hydrophilous
   (c) Both (d) None of these
5. Pusa Kiran and Pusa Kirti are the improved varieties of
   (a) Salad (b) Amaranth
   (c) Parsley (d) None of these
6. Amaranth can be grown in soil with pH as high as
   (a) 5 (b) 7
   (c) 8 (d) 10
7. Amaranth is a
   (a) Warm season crop (b) Humid season crop
   (c) Temperate season crop (d) All the above
8. The first cutting in amaranth is done
   (a) 1-2 weeks after sowing (b) 2-3 weeks after sowing
   (c) 3-4 weeks after sowing (d) None of these
9. Lettuce is a major salad crop in
   (a) South America (b) North America
   (c) Both (d) None of these
10. Non-heading type cultivar of lettuce is
    (a) Punjab Lettuce No-1 (b) Dark Green
    (c) White Boston (d) None of these
11. Great lakes and Imperial-859 are crisp head type of
    (a) Parsley (b) Lettuce
    (c) Celery (d) None of these
12. Slowbolt and Chinese Yellow are leafy cultivars of
    (a) Lettuce (b) Celery
    (c) Parsley (d) None of these
13. White Boston cultivar of lettuce is
    (a) Butter head type (b) Cos type
    (c) Leaf type (d) None of these

14. Dark green coloured cultivar of lettuce is
   (a) Butter head type (b) Cos type
   (c) Leaf type (d) None of these
15. Tip burn resistant variety of lettuce is
   (a) Great Lakes (b) Slowbolt
   (c) Chinese Yellow (d) None of these
16. Lettuce seeds don't germinate when soil temperature is above
   (a) 10°C (b) 15°C
   (c) 20°C (d) None of these
17. High temperature promotes seed stalk and causes a bitter taste in lettuce due to
   (a) Tip-burn injury (b) Shoot injury
   (c) Rent injury (d) None of these
18. Lettuce seed may be under dormancy when temperature is
   (a) 4-6°C for 3-5 days (b) 2-3°C for 10 days
   (c) 0°C for 2 days (d) None of these
19. Parsley (*Petroselinum crispum*) is rich source of
   (a) Iron (b) Vitamin C
   (c) Both (d) None of these
20. Seed rate of parsley is
   (a) 100-200 g/ha (b) 200-250 g/ha
   (c) 250-300 g/ha (d) None of these
21. Edible part of parsley is
   (a) Leafy stalks (b) Petioles
   (c) Both (d) None of these
22. *Apium graveolens* is the botanical name of
   (a) Parsley (b) Celery
   (c) Amaranth (d) None of these
23. Edible part of celery is
   (a) Leafy stalk (b) Petioles
   (c) Both (d) None of these
24. High temperature in celery leaves causes
   (a) Shoot injury (b) Root injury
   (c) Bitterness (d) None of these
25. Celery crop bolts when temperature falls below
   (a) 10°C (b) 12°C
   (c) 15°C (d) 20°C
26. Self-blanched and green leaved vegetable is
   (a) Celery (b) Parsley
   (c) Both (d) None of these

27. In celery green leaved varieties are preferred in
    (a) Chinese market (b) Indian market
    (c) Both (d) None of these
28. Self-blanched cultivar of celery is
    (a) Florida Golden (b) Golden
    (c) Both (d) None of these
29. Celery is propagated by
    (a) Seeds (b) Tuber
    (c) Cutting (d) None of these
30. In celery, blanching is done to make the
    (a) Crop crispy (b) Reduce acrid flavour
    (c) Increased tenderness (d) All the above
31. Spinach beet or palak is one of the most important vegetables in
    (a) Tropical region (b) Sub-tropical region
    (c) Both (d) None of these
32. Arka Anupma is the variety of
    (a) Amaranth (b) Celery
    (c) Spinach/*Vilayati palak* (d) None of these
33. Leafy vegetable used as pot herb is
    (a) Spinach beet (b) Spinach
    (c) Both (d) None of these
34. Leafy vegetable which contains low oxalic acid is
    (a) Spinach beet (b) Amaranth
    (c) Both (d) None of these
35. *Beta vulgaris* var. *bengalensis* is the botanical name of
    (a) Spinach beet (b) Spinach
    (c) Amaranth (d) None of these
36. Palak variety free from sourness is
    (a) Punjab Green (b) Pusa Jyoti
    (c) Both (d) None of these
37. Pusa Jyoti and All Green are the varieties of
    (a) Palak (b) Spinach
    (c) Amaranth (d) None of these
38. Palak variety Pusa Harit was developed by cross between
    (a) Sugar beet x Local palak (b) Local palak x Sugar beet
    (c) Swiss chord x Local palak (d) None of these
39. Mutant variety of palak is
    (a) Pusa Jyoti (b) Punjab Green
    (c) All Green (d) Jobner Green

40. Palak variety Pusa Harit was developed at
    (a) New Delhi (b) Katrain
    (c) Bangalore (d) None of these
41. Polyploid variety of palak is
    (a) Pusa Jyoti (b) Pusa Harit
    (c) Pusa Bharti (d) None of these
42. Spinach is known as
    (a) Spinach beet (b) Palak
    (c) Vilayati palak (d) None of these
43. Spinach crop is mainly grown in
    (a) Plains (b) Hills
    (c) Both (d) None of these
44. Prickly seeded cultivar of spinach is
    (a) Virginia Savoy (b) Khara Palak
    (c) Both (d) None of these
45. Spinach is a
    (a) Short-day plant (b) Long-day plant
    (c) Day-neutral plant (d) None of these
46. Smooth seeded cultivar of spinach is
    (a) Khara palak (b) Early Smooth Leaf
    (c) Both (d) None of these
47. Prickly seeded cultivars of spinach are best suited for
    (a) Autumn crop (b) Autumn-winter crop
    (c) Spring crop (d) Spring-summer crop
48. Smooth seeded cultivars of spinach are best suited for
    (a) Autumn crop (b) Autumn-winter crop
    (c) Spring crop (d) Spring-summer crop
49. Fenugreek is the third largest seed spice in India after
    (a) Coriander and cumin (b) Coriander and dill
    (c) Cumin and coriander (d) None of these
50. Leafy vegetable which contains dysgenic is
    (a) Fenugreek (b) Soya
    (c) Both (d) None of these
51. India is the major producer and exporter of fenugreek for
    (a) Leafy part (b) Seed
    (c) Both (d) None of these
52. Major fenugreek producing state is
    (a) Andhra pradesh (b) Uttar Pradesh
    (c) Rajasthan (d) None of these
53. *Trigonella foenum graecum* is the botanical name of
    (a) Methi (b) Kasuri Methi
    (c) Cumin (d) None of these

54. *Trigonella corniculata* is the botanical name of
    (a) Methi (b) Kasuri methi
    (c) Cumin (d) None of these

55. Slow growing type methi is
    (a) *Trigonella foenum graecum*
    (b) *Trigonella corniculata*
    (c) Both
    (d) None of these

56. Fenugreek is a
    (a) Annual (b) Biennial
    (c) Both (d) None of these

57. The flowers of fenugreek are
    (a) Leguminaceous (b) Papilionaceous
    (c) Rotate (d) None of these

58. Fruits of fenugreek are
    (a) Legumes (b) Lomentum
    (c) Pods (d) None of these

59. The flower colour of fenugreek is
    (a) White (b) Yellow
    (c) Both (d) None of these

60. Anthesis in fenugreek takes place between
    (a) 5 a.m. and 6 p.m. (b) 4 a.m. and 5 p.m.
    (c) 6 a.m. and 6 p.m. (d) None of these

61. Peak time of anthesis in fenugreek is
    (a) 9 a.m. and 6 p.m. (b) 10 a.m. and 7 p.m.
    (c) 11.30 a.m. and 12.30 p.m. (d) None of these

62. Fenugreek is a
    (a) Tropical (b) Temperate
    (c) Both (d) None of these

63. Rajendra Kranti and Hissar Sonali are the cultivars of
    (a) Methi (b) Palak
    (c) Amaranth (d) None of these

64. A spray of ascorbic acid at 250-450 ppm improves
    (a) Plant growth and enhances flowering
    (b) Number of seeds
    (c) Size of fenugreek (d) All the above

65. A spray of 10-100 ppm GA improves
    (a) Internodal length
    (b) Height and number of leaves
    (c) Leaf yield (d) All the above

66. A spray of 0.05% sodium 2,3-Isobutyrate just before flowering improves
    (a) Internodal length (b) Number of seeds
    (c) Yield of fenugreek (d) None of these

67. Seed rate of fenugreek is
    (a) 10 kg/ha (b) 15 kg/ha
    (c) 20 kg/ha (d) 25 kg/ha

68. Sowing time of fenugreek is
    (a) September (b) October
    (c) November (d) December

69. The germination period in celery seed is *(ARS, 2003)*
    (a) 10 days (b) 20 days
    (c) 30 days (d) 40 days

70. Chhoti chaulai is suitable for growing in *(ARS, 2003)*
    (a) Spring season (b) Rainy season
    (c) Both (d) None of the above

71. The flower colour of *Trigonella corniculata* (kasuri methi) is *(ARS, 2003)*
    (a) White (b) Red
    (c) Purple (d) Orange to yellow

72. The flower colour of *Trigonella foenum graccum* (common methi) is *(ARS, 2003)*
    (a) White (b) Red
    (c) Purple (d) Orange to yellow

73. Lettuce is a *(ARS, 2003)*
    (a) Cross pollinated (b) Self-pollinated
    (c) Often cross pollinated (d) None of the above

74. Great Lakes is a variety of *(ARS, 2003)*
    (a) Butter Head lettuce (b) Leaf type lettuce
    (c) Crisphead lettuce (d) None of these

75. Celery belongs to the family *(ARS, 2003)*
    (a) *Compositae* (b) *Cruciferae*
    (c) *Umbelliferae* (d) None of these

76. *Brassica pekinesis,* a useful leafy vegetable for plains and hills of North India is *(ARS, 2003)*
    (a) Chinese type (b) American type
    (c) Canadian type (d) Japanese type

77. Pusa Jyoti is a variety of *(ARS, 2003)*
    (a) Palak (b) Methi
    (c) Spinach (d) None of these

78. The most common leafy vegetable grown during summer in India is *(ARS, 2003)*
(a) Amaranth (b) Portulaca
(c) Indian spinach (d) None of these

79. Rhubarb is grown for its *(ARS, 2003)*
(a) Thick leaf stalks (b) Tubers
(c) Tender stem (d) All of these

80. Lettuce is a *(Ph.D. VGS, 2006)*
(a) Self pollinated
(b) Cross pollinated
(c) Often cross pollinated
(d) None of the above

81. Kasuri methi belongs to the species *(Ph.D. VGS, 2006)*
(a) *fenum-graecum* (b) *corniculata*
(c) *oleracea* (d) None of these

82. The leaves of spinach beet are rich source of *(Ph.D. VGS, 2006)*
(a) Vitamin A (b) Vitamin B
(c) Vitamin E (d) None of these

83. Pusa Jyoti is a variety of *(Ph.D. VGS, 2006)*
(a) Palak (b) Methi
(c) Spinach (d) None of these

84. Nursery sowing of lettuce is done in the month of *(Ph.D. VGS, 2006)*
(a) August—September (b) September—October
(c) October—November (d) December—January

85. The vegetables which are consumed in uncooked stage are known as *(ARS, 1992)*
(a) Salad crops (b) Raw vegetables
(c) Sauce crops (d) Chatni crops

86. The part of celery taken as salad is *(ARS, 1992)*
(a) Leaf stalks (b) Mid rib
(c) Stem (d) None of these

87. Parsley is a *(Ph.D. VGS, 2006)*
(a) Tuber crop (b) Bulb crop
(c) Salad crop (d) Root crop

88. Average seed yield of palak in q/ha is *(ARS, 1999)*
(a) 10 (b) 20
(c) 25 (d) None of these

89. Rhubarb and artichoke are *(ARS, 1999)*
    (a) Annual (b) Perennial
    (c) Both (d) Biennial
90. Jobner Green is a high yielding variety of *(Ph.D. VGS, 2004)*
    (a) Indian Spinach (b) Spinach
    (c) Amaranth (d) Onion
91. Leafy vegetables Palak, Amaranth and Spinach are *(Ph.D. VGS, 2004 and 2005)*
    (a) Often cross pollinated (b) Highly cross pollinated
    (c) Self pollinated (d) None of these
92. Agathi (Sesbania)is the richest source of *(Ph.D. VGS, 2005)*
    (a) Protein (b) Calcium
    (c) Both (d) None of these
93. The crop rich in folic acid is *(UP PCS (Pre), 2006)*
    (a) Spinach (b) Broccoli
    (c) Asparagus (d) Celery
94. The pulse which is also used as vegetable before podding is
    (a) Chickpea (b) Lentil
    (c) Horsegram (d) Moth bean
95. Garden beet belongs to the species *(ARS, 2003)*
    (a) *Campestris* (b) *Vulgaris*
    (c) *Sativus* (d) *Tuberosum*
96. Flowers of beet root are *(ARS, 2003)*
    (a) Staminate (b) Pistillate
    (c) Hermaphrodite (d) All of these
97. Edible part of asparagus is called *(ARS, 1999)*
    (a) Stalk (b) Bolter
    (c) Spear (d) Stem

ANSWERS

1. (c)	2. (b)	3. (a)	4. (a)	5. (b)
6. (d)	7. (d)	8. (c)	9. (b)	10. (a)
11. (b)	12. (a)	13. (a)	14. (b)	15. (a)
16. (c)	17. (a)	18. (a)	19. (c)	20. (c)
21. (c)	22. (b)	23. (a)	24. (c)	25. (c)
26. (a)	27. (b)	28. (b)	29. (a)	30. (d)
31. (c)	32. (c)	33. (c)	34. (a)	35. (a)
36. (a)	37. (a)	38. (a)	39. (d)	40. (b)

41. (c)	42. (c)	43. (b)	44. (a)	45. (b)
46. (b)	47. (b)	48. (c)	49. (a)	50. (a)
51. (b)	52. (c)	53. (a)	54. (b)	55. (a)
56. (a)	57. (b)	58. (a)	59. (c)	60. (a)
61. (c)	62. (c)	63. (a)	64. (d)	65. (d)
66. (c)	67. (d)	68. (b)	69. (c)	70. (a)
71. (d)	72. (a)	73. (b)	74. (c)	75. (c)
76. (d)	77. (a)	78. (a)	79. (a)	80. (a)
81. (b)	82. (a)	83. (a)	84. (b)	85. (a)
86. (a)	87. (c)	88. (a)	89. (b)	90. (a)
91. (b)	92. (c)	93. (a)	94. (a)	95. (c)
96. (c)	97. (c)			

CHAPTER 4

COLE CROPS

These crops belong to family *Cruciferae* which grow in cold climate. The cole crops are comprised of cabbage, cauliflower, knol-khol, brussels sprout, broccoli, kale, etc.

4.1 Cauliflower

- The first illustration and description of cauliflower was presented by the herbalist Dodens (1544).
- Cauliflower was introduced in India in 1822 by a botanist Dr. Jemson from Kew Garden, London.
- Pusa Hybrid-2 is first F_1 hybrid released by a public sector organisation.
- Generally, closer spacing is kept for early and mid season crops, while wider spacing for late cultivars is recommended.
- If the harvests of cauliflower are late, the curds start loosening because of emergence of flower stalk. They may become leafy, ricy and fuzzy at later stage.

4.2 Cabbage

- Cabbage is distinguished by its swollen heads which is formed by thickening of edible bud with thick, tightly packed and overlapping leaves, manifesting a large head.
- Cabbage has an anti-cancer property. It protects against bowel cancer due to the presence of indole-3-carbinol.
- Cabbage seed production is taken up in hilly regions of India which fulfils its chilling requirement.
- Cabbage bears a special kind of bicarpellary fruit called the 'siliqua'.
- The cultivated cabbage is a biennial plant, but its progenitor is herbaccous perennial wild cabbage.
- Self-incompatibility is found in cabbage, the flowers are protogynous.
- Pusa Drum Head variety possesses field resistance to black leg caused by *Phoma lingum.*
- Pusa Ageti is the first tropical variety developed for cultivation under high temperature conditions and which

can produce seeds under subtropical conditions of North Indian plain.

- Pusa Sambandh is a synthetic variety of cabbage.
- Cabbage produces seeds in the temperate areas only.
- Cabbage seedlings become 'Lanky' due to over application of nitrogen.
- Pusa Mukta cultivar is resistant to black rot disease.
- The inflorescence in knol-khol is racemose and has sporophytic self-incompatibility.
- In India, knol-khol is more popular in Kashmir.
- Chinese cabbage (*Brassica campestris* L.) is the most important crop grown in China.
- Chinese cabbage is grown as pot herb and also as salad crop.
- Chinese Sarson No. 1 is a non-heading type, resistant to alternaria leaf spot.

4.3 Broccoli

- The inflorescence in broccoli is cymose.
- Genic male sterility has been reported in broccoli by Cole (1959).
- Broccoli has sporophytic self-incompatibility.
- U.S.A. is the largest producer of sprouting broccoli in the world.
- Sprouting broccoli has about 130 times more vitamin A content than cauliflower and 22 times more than cabbage.
- Purple type cultivars are more hardy than white type and can withstand lower temperature.
- Green type cultivars are more commonly cultivated than white type.
- Purple Sicilian broccoli is also known as purple cauliflower.
- Palam Samridhi is mainly recommended for sub-tropical conditions.
- Early and mid types of broccoli are annual, whereas late types are biennial in nature.

4.4 Brussels sprout

- The inflorescence in Brussel's sprouts is racemose.
- Genic male sterility has been reported in Brussel's sprouts by Johnson (1958).
- Brussel's sprouts have sporophytic self-incompatibility.
- Brussel's sprouts is frost resistant and late cultivar which can withstand temperature as low as -10°C.

4.5 Kale

- Kale (*Brassica oleracea* var. *accephala*) is a minor cole crop which is propagated by seeds.
- Kale is the hardiest crop and can withstand temperature as low as -15°C.
- Dwarf Green Curled Scotch, Dwarf Moss Curled (dwarf) and Moss Curled and Hamburger Market (medium tall) are important dwarf varieties of kale.
- Karam Sag is one of the medium tall variety mostly grown in Jammu and Kashmir.
- Scotish and Siberian varieties are more suited for temperate regions.

The objective questions on these cole crops are given below :

1. *Brassica oleracea* L. var. *botrytis* is the botanical name of
   (a) Cabbage (b) Cauliflower
   (c) Knol-khol (d) None of these
2. The first illustration and description of cauliflower was presented by
   (a) Harbhajan Singh (1942) (b) Kalloo (1943)
   (c) Herbalist Dodens (1944) (d) None of these
3. Cauliflower has been descended through mutation and selection from common ancestor, the wild cabbage (*Brassica oleracea* L. var. *sylvestris*) as adopted by
   (a) Dr. Jemson (b) Dr. Kalloo
   (c) Herbalist Dodens (d) None of these
4. Cauliflower was introduced in India in 1822 by --------- from Garden, London.
   (a) Dr. Jemson (b) Dr. Kalloo
   (c) Herbalist Dodens (d) None of these
5. Pusa Hybrid-2 is the first F_1hybrid of cauliflower released by --------- organisation.
   (a) Public sector (b) Private sector
   (c) Both (d) None of these
6. Pusa Himjyoti is the only variety which may be grown during--------- in the hills
   (a) August – September (b) September– October
   (c) April – July (d) None of these
7. When tropical cultivars of cauliflower are grown at lower temperature, they produce small size curds. This is commonly termed as
   (a) Bolting (b) Buttoning
   (c) Both (d) None of these

8. Seed rate (g/ha) of cauliflower is
   (a) 400-500 (b) 500-600
   (c) 600-700 (d) None of these
9. Pusa Katki and Pusa Deepali are the improved varieties of
   (a) Cauliflower (b) Cabbage
   (c) Chinese cabbage (d) None of these
10. Seed rate (g/ha) of cabbage is
    (a) 300-400 (b) 400-500
    (c) 500-600 (d) None of these
11. Botanical name of red cabbage is
    (a) *Brassica oleracea* var. *capitata*
    (b) *Brassica oleracea* var. *rubra*
    (c) *Brassica oleracea* var. *sylvestris*
    (d) None of these
12. *Brassica oleracea* L. var. *capitata* is the botanical name of
    (a) Cabbage (b) Red Cabbage
    (c) Cauliflower (d) None of these
13. Cabbage has anti-cancer property due to
    (a) Indole-2-carbinol (b) Indole-2-butaric
    (c) Indole-3-carbind (d) None of these
14. Cabbage belongs to family
    (a) *Cruciferae* (b) *Cucurbitaceae*
    (c) Both (d) None of these
15. *Brassica olesacea* L. var. *sylvestris* is also known as
    (a) Field cabbage (b) Wild cabbage
    (c) Both (d) None of these
16. *Brassica oleracea* var. *sylvestris* is also known as
    (a) Cliff cabbage (b) Colewort
    (c) Both (d) None of these
17. Cabbage bears seed in special kind of bicarpellary fruit which is called
    (a) Pod (b) Bud
    (c) Siliqua (d) None of these
18. Cultivated cabbage is
    (a) Perennial (b) Biennial
    (c) Both (d) None of these
19. Wild cabbage is
    (a) Biennial (b) Perennial
    (c) Both (d) None of these
20. *Brassica oleracea* var. *sabauda* is the botanical name of
    (a) White cabbage (b) Red cabbage
    (c) Savoy cabbage (d) None of these

21. Self-incompatibility is found in
    (a) Cauliflower (b) Pea
    (c) Cabbage (d) None of these
22. The flowers of cabbage are
    (a) Protandrous (b) Protogynous
    (c) Both (d) None of these
23. The number of days taken by early group of cabbage to reach edible stage is
    (a) 40-50 (b) 55-70
    (c) 70-90 (d) None of these
24. Black leg is the major disease of
    (a) Cauliflower (b) Cabbage
    (c) Both (d) None of these
25. Black leg disease of cabbage is caused by
    (a) *Phoma sylivistis* (b) *Phoma lingum*
    (c) *Phoma chinensis* (d) None of these
26. *Phoma lingum* resistant variety of cabbage is
    (a) Early Drum Head (b) Pusa Drum Head
    (c) Late Drum Head (d) None of these
27. First tropical variety of cabbage is
    (a) Pusa Ageti (b) Pusa Sambandh
    (c) Both (d) None of these
28. Cabbage variety which can be cultivated under high temperature conditions is
    (a) Pusa Drum Head (b) Pusa Ageti
    (c) Pusa Sambandh (d) None of these
29. The conditions where the seed of Pusa Ageti can be produced are
    (a) Tropical (b) Sub-tropical
    (c) Temperate (d) None of these
30. Pusa Sambandh is a synthetic variety of
    (a) Cauliflower (b) Cabbage
    (c) Chinese cabbage (d) None of these
31. Cabbage grown in saline soils is more susceptible to disease
    (a) Hollow heart (b) Black-neck
    (c) Black-leg (d) None of these
32. Cabbage produces seed only in
    (a) Tropical areas (b) Temperate areas
    (c) Sub-tropical areas (d) None of these

33. Cabbage may be grown almost throughout the year in
   (a) Mid hills (b) Southern hills
   (c) Both (d) None of these
34. Cabbage seedlings become 'Lanky' due to over dose of
   (a) Phosphorus (b) Nitrogen
   (c) Calcium (d) Sulphur
35. Cabbage seedlings are transplanted at
   (a) 2-4 true leaf stage (b) 4-6 true leaf stage
   (c) 6-8 true leaf stage (d) None of these
36. Black rot disease resistant variety of cabbage is
   (a) Pride of India (b) Golden Acre
   (c) Pusa Mukta (d) Copen Hagen Market
37. *Brassica oleracea* var *gongyloides* or *caulorapa* is the botanical name of
   (a) Cabbage (b) Cauliflower
   (c) Knol- Khol (d) None of these
38. Knol-Khol is also called as
   (a) Broccoli (b) Kohlrabi
   (c) Both (d) None of these
39. Seed rate (g/ha) of Kohlrabi is
   (a) 400-500 (b) 500-600
   (c) 600-700 (d) 1000-1500
40. The inflorescence in Knol-Khol is
   (a) Cymose (b) Racemose
   (c) Both (d) None of these
41. Sporophytic self- incompatibility is found in
   (a) Cabbage (b) Knol-Khol
   (c) Both (d) None of these
42. Edible part of Knol-Khol is (ARS, 1999)
   (a) Root (b) Leaves
   (c) Swollen stem (d) None of these
43. White Vienna and King of Market are improved varieties of
   (a) Knol-khol (b) Cabbage
   (c) Cauliflower (d) None of these
44. Chromosome number in Chinese cabbage is
   (a) 2n = 18 (b) 2n = 20
   (c) 2n = 24 (d) None of these
45. Chromosome number in cauliflower is
   (a) 2n = 18 (b) 2n = 20
   (c) 2n = 24 (d) None of these

46. Botanical name of Chinese cabbage is
    (a) *Brassica compestris* L. var. *caulorapa*
    (b) *Brassica compestris* L. var. *longifora*
    (c) *Brassica compestris* L. var. *Chinensis*
    (d) None of these
47. Seed rate/(g/ha) of chinese cabbage is
    (a) 100 g/ha (b) 200 g/ha
    (c) 400 g/ha (d) 500 g/ha
48. Seed rate of Chinese cabbage (kg/ha) for direct sowing is
    (a) 1.0 (b) 2.0
    (c) 2.5 (d) 3.0
49. Chinese cabbage is grown as pot herb and also as
    (a) Leafy crop (b) Salad crop
    (c) Bulb crop (d) None of these
50. Chinese cabbage originated from hybridization between turnip and
    (a) Kole (b) Chinese cabbage/Pak-choi
    (c) Both (d) None of these
51. Chinese sarson No-1 is .... type of chinese cabbage.
    (a) Heading (b) Non-heading
    (c) Both (d) None of these
52. *Brassica oleracea* L. var. *italica* is the botanical name of
    (a) Chinese cabbage (b) White cabbage
    (c) Cabbage (d) Broccoli
53. Chromosome number in broccoli is
    (a) 2n = 16 (b) 2n = 18
    (c) 2n = 20 (d) None of these
54. The infloresence of broccoli is
    (a) Cymose (b) Racemose
    (c) Both (d) None of these
55. Genetic male sterlity has been reported in broccoli by
    (a) Alter (1959) (b) Cole (1959)
    (c) Both (d) None of these
56. The type of incompatibility in broccoli is
    (a) Gametophytic (b) Sporophytic
    (c) Both (d) None of these
57. The country with largest production of broccoli in the world is
    (a) China (b) U.S.A.
    (c) India (d) None of these

58. Seed rate (g/ha) of broccoli is
    (a) 100-200 (b) 200-300
    (c) 300-400 (d) 400-500

59. Edible part of broccoli is
    (a) Green buds (b) Thick fleshy floral stalk
    (c) Both (d) None of these

60. Sprouting broccoli is the richest source of
    (a) Sulphoraphane (b) Sulpharadene
    (c) Both (d) None of these

61. Anticancerous property is found in
    (a) Cabbage (b) Cauliflower
    (c) Broccoli (d) None of these

62. Purple Sicilian broccoli is also known as
    (a) Purple cabbage (b) Purple cauliflower
    (c) Chinese cabbage (d) None of these

63. Palam Samridhi is an improved variety of
    (a) Broccoli (b) Cabbage
    (c) Cauliflower (d) None of these

64. The broccoli grown in India are susceptible to frost are
    (a) Early and mid types (b) Mid and late types
    (c) Only late types (d) None of these

65. Early and mid types of broccoli are annual but late types are
    (a) Annual (b) Biennial
    (c) Both (d) None of these

66. Which type of broccoli cultivars require vernalization
    (a) Early (b) Mid
    (c) Late (d) None of these

67. *Brassica oleracea* L. var. *gemmifera* is the botanical name of
    (a) Sprouting broccoli (b) Brussels sprouts
    (c) Kale (d) Cauliflower

68. Seed rate (g/ha) of brussels sprouts is
    (a) 200-300 (b) 300-400
    (c) 400-500 (d) 500-600

69. The inflorescence of brussels sprouts is
    (a) Cymose (b) Racemose
    (c) Both (d) None of these

70. Genetic male sterility has been reported in brussels sprouts by
    (a) Cole (1959) (b) Johnson (1958)
    (c) Both (d) None of these

71. Sporophytic self incompatibility is found in
(a) Broccoli (b) Brussels sprouts
(c) Knol-Khol (d) All the above

72. Edible part of brussels sprouts and cabbage is
(a) Swollen axillary buds (b) Buds
(c) Green buds and thick fleshy floral stalk
(d) None of these

73. Brussels sprout is also known as
(a) Sprouts (b) Buttons
(c) Mini cabbage (d) All the above

74. Brussels sprouts is a
(a) Cool crop (b) Moisture loving crop
(c) Both (d) None of these

75. Brussels sprouts is frost resistant and late cultivar can with-stand temperature as low as ---------°C
(a) 10 (b) -10
(c) 15 (d) -15

76. *Brassica oleracea* L. var. *acephala* is the botanical name of
(a) Cabbage (b) Kale
(c) Broccoli (d) None of these

77. Kale is propagated through
(a) Seed (b) Vegetable
(c) Both (d) None of these

78. Karam Sag and Dwarf Green Curled Scotch are the improved varieties of
(a) Cabbage (b) Kale
(c) Cauliflower (d) None of these

79. Seed rate (g/ha) of kale is
(a) 300-400 (b) 400-500
(c) 500-600 (d) None of these

80. Purple Vienna is a.......... variety of knol-khol. *(ARS, 2003)*
(a) Early (b) Mid season
(c) Late (d) All season

81. 'Dania' is a variety of *(ARS, 2003)*
(a) Cabbage (b) Cauliflower
(c) Knol-khol (d) None of these

82. Round head type of cabbage cultivars are of .... maturity type. *(ARS, 2003)*
(a) Early (b) Mid
(c) Late (d) None of these

83. Which of the following is an early variety of cauliflower? *(ARS, 2003)*
(a) Pusa Ketaki (b) Synthetic Second
(c) Snowball-16 (d) None of these

84. Which of the following is the early variety of cabbage? *(ARS, 2003)*
(a) Pride of India (b) Early Drum Head
(c) Chieftain (d) None of these

85. Whiptail in cauliflower is caused by *(ARS, 2003, UP PCS (Pre), 2006)*
(a) Excess of boron (b) Deficiency of boron
(c) Excess of molybdenum (d) Deficiency of molybdenum

86. White Vienna is......... season variety of knol-khol. *(ARS, 2003)*
(a) Early (b) Mid
(c) Late (d) All season

87. Fleshy edible part of knol-khol is an enlargement of *(ARS, 2003)*
(a) Root (b) Stem
(c) Leaf (d) None of these

88. Early cauliflower is ready for harvesting in the period of *(ARS, 2003)*
(a) Mid-September to mid-November
(b) Mid-November to mid-January
(c) Mid-January to April (d) None of the above

89. Hardening of cabbage seedlings may be done by with-holding irrigation prior to transplanting *(ARS, 2003)*
(a) 4-6 days (b) 6-8 days
(c) 8-10 days (d) 10-12 days

90. Healthy and stocky seedlings of cauliflower are transplanted at a distance of *(ARS, 2003)*
(a) 60 x 30-45 cm (b) 60 x 60 cm
(c) 45 x 45 cm (d) 40 x 15 cm

91. The most nutritive type of broccoli is *(ARS, 2003)*
(a) Purple coloured (b) White coloured
(c) Green coloured (d) None of these

92. The first synthetic cabbage variety was recommended in the year *(ARS, 1992)*
(a) 1990 (b) 1991
(c) 1992 (d) 1993

93. Browning in cauliflower can be corrected by the application of *(UPPCS (Pre), 1994)*
(a) Ammonium molybdenum (b) Boric acid
(c) Zinc sulphate (d) Urea

94. Which of the following varieties of cauliflower is self blanching type ? *(Ph.D. VGS, 2006)*
(a) Pant Shunhra (b) Snow Ball-16
(c) Hissar (d) Pusa Ketki

95. The best suited soil pH range for cauliflower is *(RAS (Pre), 2003)*
(a) 5.5-6.5 (b) 6.5-7.5
(c) 7.5-8.5 (d) 6.0-7.0

96. When the surface of curd in cauliflower is loose and has velvety appearance then it is called *(RAS (Pre), 2003)*
(a) Fuzziness (b) Blindness
(c) Riciness (d) Browning

97. Vegetable crops belonging to cole crops group thrive best in *(Ph.D. VGS, 2006)*
(a) Cool weather (b) Hot humid climate
(c) Arid climate (d) Semi-arid climate

98. All cole crops *i.e.* cabbage, cauliflower, knol-khol or kohlrabi, broccoli, Brussels sprout and Chinese cabbage have developed from wild cabbage known as *(Ph.D. VGS, 2006)*
(a) Cole-wart (b) Cole-sprout
(c) Cole-coli (d) None of these

99. Early cauliflower varieties are available for consumption in *(Ph.D. VGS, 2006)*
(a) December to January
(b) Mid September to mid October
(c) January to February (d) March to April

100. Mid season cauliflower varieties are available for consumption in *(Ph.D. VGS, 2006)*
(a) Mid-November to mid-December
(b) Mid-September to mid-October
(c) Mid-January to mid-February (d) April to May

101. Which one of the following is a cole crop ? *(UP PCS (Pre), 2004, 2005)*
(a) Beet root (b) Turnip
(c) Knol-khol (d) Spinach

102. Isolation distance of cabbage for foundation seed production is *(ARS 1999, Ph.D VGS, 2004)*
(a) 800 m (b) 1000 m
(c) 1600 m (d) None of these

103. Riciness in cauliflower is due to *(ARS, 1999)*
   (a) High temperature (b) Low temperature
   (c) Both (d) None of these

104. Indian cauliflower is tolerant to *(ARS, 1999)*
   (a) Low temperature (b) High temperature
   (c) Both (d) None of these

105. The edible portion of cabbage is *(ARS, 1999)*
   (a) Leaf (b) Head
   (c) Root (d) Stem

106. Kunwari is the variety of *(ARS, 1999)*
   (a) Knol-khol (b) Cabbage
   (c) Cauliflower (d) None of these

107. The seed required for growing seedlings for planting in one hectare area of Knol-khol is *(ARS, 1999)*
   (a) 500-800 g (b) 900-1000 g
   (c) 1000-1500 g (d) None of these

108. The place of origin of sprouting broccoli is *(ARS, 1999)*
   (a) Italy (b) India
   (c) America (d) None of these

109. Brussels sprouts matures in *(ARS, 1999)*
   (a) 90-100 days (b) 60-80 days
   (c) 120-180 days (d) 40-60 days

110. The center of origin of cauliflower is *(ARS, 1999)*
   (a) Germany (b) France
   (c) Greece (d) Cyperus

111. *Brassica oleracea* var. *gongyloides* is the botanical name of *(ARS, 1999)*
   (a) Cabbage (b) Cauliflower
   (c) Knol-khol (d) None of these

112. Blindness in cauliflower is due to *(ARS, 1999)*
   (a) Excess of nitrogen (b) Nitrogen deficiency
   (c) Insect injury to terminal bud
   (d) High temperature

113. Which of the cauliflower requires chilling temperature for seed production ? *(ARS, 1999)*
   (a) Pusa Deepali (b) Pusa Synthetic
   (c) Pusa Shubhra (d) None of these

114. Late cauliflower does not set seed in plains due to absence of *(Ph.D. VGS, 2004)*
   (a) Long day and high temperature
   (b) Long day
   (c) Short day and high temperature
   (d) Short day and low temperature

115. Pusa Mukta, an improved variety of cabbage, is resistant to *(Ph.D. VGS, 2004)*
    (a) White rot (b) Black rot
    (c) Downy mildew (d) Powdry mildew

116. Cabbage is cross pollinated owing to *(Ph.D. VGS, 2004)*
    (a) Floral morphology (b) Protandry
    (c) Self-incompatibility (d) None of these

117. Bt cabbage is resistant to *(Ph.D. VGS, 2004)*
    (a) DMB (b) Cabbage butterflies
    (c) Aphid (d) Painted bug

118. The cauliflower mosaic virus particles contain *(Ph.D. VGS, 2005)*
    (a) RNA (b) DNA
    (c) Both (d) RNA or DNA

119. Cross pollination in cabbage is due to *(UPPCS (Pre), 2006*
    (a) Floral structure (b) Protandry
    (c) Protogyny (d) Self-incompatibility

120. Hollow stem of cauliflower is due to deficiency of *(UPPCS (Pre), 2006)*
    (a) Boron (b) Zinc
    (c) Molybdenum (d) Copper

121. Seed rate of sprouting broccoli (g/ha) is *(UPPCS (Pre), 2006)*
    (a) 150-250 (b) 350-400
    (c) 500-600 (d) 700-800

122. Among the following European vegetables which one is rich source of protein ? *(UPPCS (Pre), 2006)*
    (a) Broccoli (b) Brussels
    (c) Kale (d) Red cabbage

123. Flat head or drum head and savoy types of cabbage cultivars have..... maturity. *(ARS, 2003)*
    (a) Early type (b) Mid type
    (c) Late type (d) None of these

ANSWERS

1. (b)	2. (c)	3. (a)	4. (a)	5. (a)
6. (c)	7. (b)	8. (a)	9. (a)	10. (c)
11. (b)	12. (a)	13. (a)	14. (a)	15. (b)
16. (c)	17. (c)	18. (b)	19. (b)	20. (c)
21. (c)	22. (b)	23. (b)	24. (b)	25. (b)
26. (b)	27. (a)	28. (b)	29. (b)	30. (b)
31. (c)	32. (b)	33. (c)	34. (b)	35. (b)
36. (c)	37. (c)	38. (c)	39. (d)	40. (b)

41. (b)	42. (c)	43. (a)	44. (b)	45. (a)
46. (c)	47. (d)	48. (c)	49. (b)	50. (b)
51. (b)	52. (d)	53. (b)	54. (a)	55. (b)
56. (b)	57. (b)	58. (c)	59. (c)	60. (a)
61. (c)	62. (b)	63. (a)	64. (a)	65. (b)
66. (c)	67. (b)	68. (b)	69. (b)	70. (b)
71. (d)	72. (a)	73. (d)	74. (c)	75. (b)
76. (b)	77. (a)	78. (b)	79. (a)	80. (c)
81. (a)	82. (a)	83. (a)	84. (a)	85. (d)
86. (a)	87. (b)	88. (a)	89. (a)	90. (a)
91. (c)	92. (c)	93. (b)	94. (b)	95. (d)
96. (a)	97. (a)	98. (a)	99. (b)	100. (a)
101. (c)	102. (c)	103. (b)	104. (b)	105. (b)
106. (c)	107. (c)	108. (a)	109. (a)	110. (d)
111. (c)	112. (c)	113. (c)	114. (d)	115. (b)
116. (c)	117. (a)	118. (b)	119. (d)	120. (a)
121. (b)	122. (c)	123. (c)		

CHAPTER 5

CUCURBITACEOUS VEGETABLES

This is a very large group of vegetables belonging to the family *Cucurbitaceae*. The most of vegetables are hot season few of them (water-melon, muskmelon, cucumber, long melon) are most popular fruits.

5.1 Cucumber

- The immature fruits of cucumber (*Cucumis sativus* L.) (*gherkin*) are used as *salad* and for pickling.
- Cucumber is the second most widely cultivated cucurbit after watermelon.
- The open pollinated cultivars of cucumber are monoecious in sex form.
- The first F_1, hybrid, Pusa Sanyog, developed in India in 1971, has a gynoecious female parent.
- Generally slicing cucumber has white spine colour and pickling cucumber has black spine colour.
- Variety, Himangi is resistant to bronzing.
- Pusa Sanyog is an F_1 hybrid of Japanese Gynoecious line x Green Long of Naples.
- Poinsett is resistant to downy mildew, powdery mildew, anthracnose and angular leaf spot.
- Chilling injury as a physiological storage disorder of cucumber is noticed when fruits are exposed to temperature below-10°C for long period.

5.2 Pumpkin

- Pumpkins have high carotene content and good keeping quality.
- The flowers of pumpkin are more nutritious than fruits.
- Flowers of pumpkin are unisexual, solitary or fasciculate and are lemon-yellow to deep orange in colour.
- Pumpkin is highly cross pollinated due to monoecious nature and entomophily.
- Arka Chandan is very rich in carotene content. (3331 iu/100 g of flesh).

- Pusa Vikas is most suitable for cultivation in spring-summer season in North India.
- Short-day, low-night temperature and high relative humidity are best for pumpkin production.
- In pumpkin, high nitrogen under temperate condition promotes maleness in flowering resulting in low fruit set and low yield.

5.3 Water-melon

- A seedless variety of watermelon, namely Pusa Bedana is an F_1 hybrid between tetra 2 x Pusa Rasal.
- Cool nights and warm days are ideal for accumulation of sugar in water-melon.
- The average sweetness in water-melon is recorded around 9-10% TSS.

5.4 Musk-melon

- Mature fruits of musk-melon (*Cucumis melo* L.) are round in shape and 8-16 cm in diameter.
- Anthesis in musk-melon takes place at 5.30-6.30 a.m. at temperature of 22-29°C.
- Keeping quality of Arka Rajhans is excellent.
- Melons require tropical climate and fairly high temperatures of 35-40°C during fruit development.
- Cool nights and warm days are ideal for accumulation of sugar in the fruit.
- In musk-melon, application of ethrel (250 ppm) increases the fruiting and yield.
- Musk-melon is a climacteric fruit.
- The fruits, when mature, slip out easily from the vine with a little pressure or jerk or if not, get separated the next day. This is called full slip stage. Most of the musk-melon cultivars behave in this way.

5.5 Bottle gourd

- Bottle gourd has great export potential. The fruits exported should be light green to dark green in colour having a length of 30-40 cm and should be straight.
- Pusa Meghdoot is F_1 hybrid between Pusa Summer Prolific Long and Sel 2.
- Pusa Manjari is a round fruited F_1 hybrid between Pusa Summer Prolific Round and Sel 11.

- Pusa Summer Prolific Round is prolific bearer and heavy yielder.
- Fruits of Arka Bahar are devoid of crook-neck and have flesh with pleasant aroma.
- Pusa Hybrid 3 is very early in maturity.
- The Samrat cultivar of bottle gourd has the highest yield potential on bower system.
- Maleic hydrazide at 400 ppm promotes the female flower production and increases fruit set and in turn the yield.

5.6 Ash gourd

- Ash gourd is an extensive trailing or climbing annual herb.
- Ash gourd or wax gourd or white gourd or Chinese preserving melon is also known as *Petha kaddu.*
- Mudliar is a variety of ash gourd.
- Short-day, low-night temperature and humid climate are good for production of female flowers in ash gourd.

5.7 Sponge gourd and ridge gourd

- *Luffa cylindrica* M. J. Roem. (Synonym : *L. aegyptica* Mill.) (sponge gourd, smooth loofah, vegetable sponge, dish cloth gourd) is an annual monoecious climber.
- The flowers of sponge gourd open in early morning hours (4.0-8.0 am).
- The sponge gourd fruits contain higher protein and carotene than ridge gourd.
- Sponge gourd and ridge gourd contain a gelatinous compound called *luffein.*
- Pusa Chikni, Phule Prajakta and Pusa Supriya are the varieties of sponge gourd.
- *Luffa acutengula* (L.) Roxb. (angled loofah, ridge gourd, ribbed gourd) is an annual monoecious climber.
- The fruits of ridge gourd are 14-15 cm long and club shaped. The surface is distinctly ribbed.
- Satputia is a cultivar of Bihar, which is hermaphrodite in sex form and produces smaller fruits in clusters.
- Pusa Nasdar, Konkan Harita, and Punjab Sadabahar are the varieties of ridge gourd.

5.8 Parwal

- Pointed gourd is commonly called *parwal* and it is a perennial cucurbit.
- Bengal-Assam area is the primary centre of origin of pointed gourd.

- The fruit of pointed gourd is fusiform and 8-12 cm long.
- In order to plant one hectare of pointed gourd, about 2,000-2,500 cuttings are required.
- Pointed gourd is vegetatively propagated through vine cuttings and root suckers.
- Training the pointed gourd crop gives high yields.
- Seed propagation is avoided in pointed gourd because of poor seed germination and since it is dioecious in nature, it produces male and female plants in equal proportion, if they are planted from the seeds.

5.9 Ivy gourd

- Ivy gourd is also known as Coccinia or little gourd or scarlet gourd.
- Ivy gourd is most commonly used as a vegetable in southern and central India.
- The fruits of Ivy gourd are rich source of carbohydrates, proteins, vitamins A and C.
- Ivy gourd is dioecious in nature.
- Ivy gourd is a climbing perennial and has tuberous roots.
- Ivy gourd is propagated through stem cuttings.
- Ivy gourd fruits are produced almost throughout the year.
- An average of 200-300 fruits weighing 3-4 kg can be harvested per vine per year.

5.10 Snake gourd

- Konkan Sweta is a variety of snake gourd.
- Snake gourd needs a warm and humid climate for best growth. High humidity is favourable for growth and fruit development.

5.11 Bitter gourd

- Bitter gourd is also known as balsampear or bitter cucumber.
- Bitter gourd fruits are rich in iron.
- The cucurbitacin (bitter glucocide) may help in preventing spoilage of cooked vegetable of bitter gourd.
- Anthesis and anther dehiscence in bitter gourd occurs early in the morning.
- White coloured varieties of bitter gourd are less bitter in taste and preferred in south India.
- Phule Green is developed by cross between Green Long and Delhi Local.

5.12 Round gourd

- Round gourd is also known as Indian squash or round melon or tinda or squash melon.
- The fruit has cooling effect and contains vitamin A.
- Arka Tinda is an early summer season cultivar.
- The optimum temperature for seed germination of round gourd is 27°C.

5.13 Winter squash

- Winter squash (*vilayati kaddu, halwa kaddu*) is a monoecious, annual climber.
- Arka Suryamukhi is an improved variety. It is resistant to fruit fly.
- Winter squash is commonly grown in the hills.
- Pink Banana is a variety of winter squash.
- Summer squash is also called vegetable marrow or field pumpkin.
- Pusa Alankar is an F_1 hybrid between EC 207050 and Sel. 1.
- Patty PAN, an introduction from USA, is a short-duration variety.

5.14 Long-melon

- Long-melon (*kakri or tar*) is more commonly grown in north India.
- Long-melon can tolerate cool climate better than muskmelon and snap melon.
- Arka Sheetal is an improved cultivar of long-melon.
- Long-melon fruits are harvested when they are very tender and at about half-grown stage.
- Green type of long-melon fruits are comparatively good for transportation.

5.15 Snap-melon

- Botanically, snap melon is *Cucumis melo* var. *mamordica*.
- Snap-melon (*Phoot* or *Kachra*) is not as popular as long melon but it is still grown during summer and rainy season in north India.
- Usually the fruits burst when the snap-melon is ripe, therefore, it is popularly called *phoot*.

5.16 Chow-chow

- Chow-chow or cristophine chow-chow or chayote or asks or choco (*Sechium edule* **Swartz**) is a herbaceous perennial climbing monoecious vine with large tuberous roots.
- Chow-chow is grown principally for its pear-shaped fruits, which are cooked in many ways.
- The fruits of chow-chow are fleshy and pyriform with longitudinal furrows.
- Chow-chow shows vaviparous germination.
- Chayote is the only cultivated member of *Cucurbitaceae* family which is single-seeded.

5.17 Spine-gourd

- Kakrol or Spine gourd (*Momordica dioica*) is a large, dioecious perennial climber.
- Kakrol roots are tuberous.
- Kakrol is rich in protein and vitamin C.
- Kakrol is generally propagated by its tuberous roots.
- It can also be propagated by seeds.
- About 10% of male plants should be planted for good fruit set.
- Kakrol seeds contain oil which is used as an illuminant.

The objective type questions on *Cucurbetaceous* vegetables are given below :

1. *Momordica dioica* is the botanical name of
   (a) Kakrol or Spine gourd (b) Sponge gourd
   (c) Bottle gourd (d) None of these
2. The young leaves are also consumed as leafy vegetable in
   (a) Cucumber (b) Spine gourd
   (c) Bitter gourd (d) None of these
3. Spine gourd is rich in
   (a) Protein (b) Vitamin-C
   (c) Both (d) None of these
4. Spine gourd is generally propagated through
   (a) Seeds (b) Tuberous roots
   (c) Both (d) None of these
5. Seed rate of spine gourd is
   (a) 1-2 kg/ha (b) 2-3 kg/ha
   (c) 3-4 kg/ha (d) None of these

6. The female and male plant ratio for planting of spine gourd is
   (a) 10 : 1 (b) 1 : 10
   (c) both (d) None of these
7. Kakrol seeds contain oil which is used as
   (a) Illuminant (b) Hypolitic
   (c) Both (d) None of these
8. *Momordica cochinchinensis* is the botanical name of
   (a) Spine gourd (b) Kartoli/Sweet gourd
   (c) Sponge gourd (d) None of these
9. Sweet gourd is a perennial plant but its vines are
   (a) Perennial (b) Annual
   (c) Both (d) None of these
10. Sweet gourd goes in dormancy after producing fruits in the month of
    (a) September to October (b) October to November
    (c) November to December (d) None of these
11. Sweet gourd is propagated through
    (a) Tuberous roots (b) Seeds
    (c) Both (d) None of these
12. *Sechium edule* swartz is the botanical name of
    (a) Sponge gourd (b) Spine gourd
    (c) Sweet gourd (d) Chow-chow
13. Chayote is a
    (a) Herbaceous plant (b) Perennial plant
    (c) Monoecious plant (d) All the above
14. Edible part of chow-chow is
    (a) Seed (b) Tender fruits
    (c) Root (d) None of these
15. Chayote is propagated through
    (a) Perennial vines (b) Annual vines
    (c) Both (d) None of these
16. Chayote is a perennial crop but its vines are
    (a) Perennial (b) Annual
    (c) Both (d) None of these
17. The germination in chow-chow is
    (a) Epigeal (b) Hypogeal
    (c) Vaviparous (d) None of these
18. Single seeded cucurbitaceous vegetable is *(Ph.D. VGS, 2005)*
    (a) Cucumber (b) Spine gourd
    (c) Chayote (d) None of these

19. *Cucumis melo* var. *momordica* is the botanical name of
    (a) Snake gourd (b) Snap-melon (*Phoot*)
    (c) Long-melon (d) None of these
20. *Cucumis melo* var. *utilissimus* is the botanical name of
    (a) Snake gourd (b) Snap-melon
    (c) Long-melon (d) None of these
21. Arka Sheetal is an improved variety of
    (a) Long-melon (b) Snap-melon
    (c) Snake-melon (d) None of these
22. Long-melon fruits are picked at
    (a) Half- grown stage (b) Full- grown stage
    (c) Both (d) None of these
23. *Cucurbita moschata* is the botanical name of
    (a) Pumpkin (b) Winter squash
    (c) Summer squash (d) None of these
24. *Cucurbita maxima* is the botanical name of
    (a) Pumpkin (b) Winter squash
    (c) Summer squash (d) None of these
25. *Cucurbita pepo* L. is the botanical name of
    (a) Pumpkin (b) Winter squash
    (c) Summer squash (d) None of these
26. Winter squash is also known as
    (a) Vilayati kaddu (b) Chhappan kaddu
    (c) Both (d) None of these
27. Arka Suryamukhi is an improved variety of
    (a) Winter squash (b) Summer squash
    (c) Both (d) None of these
28. The variety Arka Suryamukhi is resistant to
    (a) Fruit borer (b) Shoot fly
    (c) Fruit fly (d) None of these
29. Winter squash is commonly grown in
    (a) Plains (b) Hills
    (c) Ice point (d) None of these
30. Pink Banana is the variety of
    (a) Chilli (b) Tomato
    (c) Winter squash (d) None of these
31. Summer Squash is also called
    (a) Vegetable marrow (b) Field pumpkin
    (c) Chhappan Kaddu (d) All the above

32. Pusa Alankar is a variety of
    (a) Winter Squash (b) Summer Squash
    (c) Both (d) None of these
33. Pusa Alankar is an F_1 hybrid between
    (a) EC 207050 x Sel.-1 (b) EC 71710 x Sel.-1
    (c) EC 207050 x Sel.-2 (d) None of these
34. Downy mildew and red pumpkin beetle resistant variety of Summer Squash is
    (a) Punjab Selection (b) Punjab Chhappan Kaddu
    (c) Both (d) None of these
35. Short duration introduced variety of summer squash is
    (a) Patty PAN (b) Pusa Alankar
    (c) Both (d) None of these
36. The fruits of summer squash are harvested at the maturity stage of
    (a) One-fourth (b) One-third
    (c) Two-third (d) None of these
37. *Praecitrullas fistulosus* Pang. is the botanical name of
    (a) Round gourd (b) Snake gourd
    (c) Long-melon (d) None of these
38. Round gourd is also known as
    (a) Indian squash (b) Tinda
    (c) Squash-melon (d) All the above
39. Indian squash fruit contains
    (a) Vitamin-A (b) Cooling effect
    (c) Both (d) None of these
40. Arka Tinda is a cultivar of round gourd suitable for
    (a) Early Summer Season (b) Late Summer Season
    (c) Both (d) None of these
41. The optimum temperature for seed germination of round gourd is
    (a) 10°C (b) 15°C
    (c) 20°C (d) 27°C
42. Application of MH 50 ppm at 2-4 leaf stage increases the
    (a) Number of female flowers (b) Fruit yield
    (c) Both (d) None of these
43. Seed rate (kg/ha) of Indian squash is
    (a) 2-3 (b) 3-4
    (c) 4-5 (d) None of these

44. *Momordica charantia* L. is the botanical name of
    (a) Bitter gourd (b) Sponge gourd
    (c) Indian squash (d) None of these
45. Bitter gourd is also known as
    (a) Balsampear (b) Bitter cucumber
    (c) Both (d) None of these
46. Bitter gourd fruits are rich in
    (a) Iron (b) Vitamin-B
    (c) Both (d) None of these
47. Bitterness in bitter gourd is due to
    (a) Solanine (b) Cucurbitacin
    (c) β- Carotene (d) Momordicin
48. The time of anthesis and anther dehisence in bitter gourd is
    (a) Early morning (b) Evening
    (c) Afternoon (d) None of these
49. *Momordica charantia* L. is incompatible with
    (a) *M. dioica* and *M.balsamin* (b) *M. dioica* and *M. Indica*
    (c) Both (d) None of these
50. The variety of bitter gourd which is less bitter in taste and preferred in south India is
    (a) Green (b) White
    (c) White and Green (d) None of these
51. The glucocide helps in preventing spoilage of cooked vegetable of bitter gourd is
    (a) Solanine (b) Cucurbitacin
    (c) Momordicin (d) None of these
52. Best temperature for producing female flowers in bitter gourd is
    (a) 10°C (b) 15°C
    (c) 20°C (d) None of these
53. Best temperature for seed germination of bitter gourd is
    (a) >10°C (b) >12°C
    (c) >15°C (d) None of these
54. Application of MH 50-150 ppm and Cycocel 50-100 ppm in bitter gourd increases
    (a) Vegetative growth (b) Sex ratio
    (c) Both (d) None of these
55. Ethrel increases female flowers in bitter gourd at the concentration of
    (a) 10 ppm (b) 20 ppm
    (c) 25 ppm (d) None of these

56. Pusa Vishesh and Priya are the varieties of
    (a) Bottle gourd (b) Bitter gourd
    (c) Sponge gourd (d) None of these
57. Kalyanpur Sona is the improved variety of bitter gourd suitable for
    (a) Summer season (b) Winter season
    (c) Both (d) None of these
58. Kalyanpur Baramasi is the improved variety of bitter gourd suitable for
    (a) Summer season (b) Rainy season
    (c) Both (d) None of these
59. *Trichosanthes anguina* L. is the botanical name of
    (a) Snake gourd (b) Snap gourd
    (c) Sponge gourd (d) None of these
60. Snake gourd is also known as
    (a) Sponge gourd (b) Snap gourd
    (c) Chichinda (d) None of these
61. Konkan Sweta is a variety of
    (a) Snake gourd (b) Sponge gourd
    (c) Snap gourd (d) None of these
62. Snake gourd requires
    (a) Warm climate (b) Humid climate
    (c) Both (d) None of these
63. *Coccinia cordifolia* is the botanical name of
    (a) Ivy gourd (b) Pointed gourd
    (c) Snake gourd (d) None of these
64. Ivy gourd is also known as
    (a) *Coccinia* (b) Little gourd
    (c) Scarlet gourd (d) All the above
65. Fruits of Ivy gourd are rich source of
    (a) Carbohydrates (b) Proteins
    (c) Vitamins A and C (d) Minerals
66. Ivy gourd in respect of distribution of sex is
    (a) Monoecious (b) Dioecious
    (c) Both (d) None of these
67. Ivy gourd is propagated through
    (a) Seeds (b) Stem cuttings
    (c) Tuber cutting (d) None of these
68. To get good yield of Ivy gourd the vines of male type should be planted in the field in the percentage of
    (a) 5 (b) 10
    (c) 15 (d) None of these

69. Average number of fruits of Ivy gourd harvested per vine per year is
    (a) 50-100 (b) 100-200
    (c) 200-300 (d) None of these
70. An average weight of 200-300 fruits in Ivy gourd is
    (a) 1-2 kg (b) 2-3 kg
    (c) 3-4 kg (d) 4-5 kg
71. *Tricosanthes dioica* Roxb. is the botanical name of
    (a) Ivy gourd (b) Pointed gourd
    (c) Snake gourd (d) None of these
72. Pointed gourd is also called as
    (a) Sweet gourd (b) Parwal
    (c) Both (d) None of these
73. Center of origin of pointed gourd is
    (a) India (b) Africa
    (c) China (d) None of these
74. To plant one hectare of pointed gourd number of cuttings required is
    (a) 2000-2500 (b) 2500-3000
    (c) 3000-5000 (d) 5000-8000
75. Pointed gourd is vegetatively propagated through
    (a) Vine cuttings (b) Root cuttings
    (c) Both (d) None of these
76. To obtain good yield of pointed gourd, vines of male type should be planted in the field in the percentage of
    (a) 5 (b) 12
    (c) 20 (d) None of these
77. During winter pointed gourd becomes dormant and sprouts again in
    (a) Winter (b) Rainy
    (c) Summer (d) None of these
78. Swarna Rekha and Swarna Anupma are the improved varieties of
    (a) Ivy gourd (b) Pointed gourd
    (c) Snake gourd (d) None of these
79. *Cucumis sativus* L. is the botanical name of
    (a) Cucumber (b) Gherkin
    (c) Both (d) None of these
80. *Cucumis anguria* L. is the botanical name of
    (a) Cucumber (b) Gherkin
    (c) Both (d) None of these

81. *Cucumis anguria* L. is also known as
    (a) Pickling cucumber (b) Wild cucumber
    (c) Both (d) None of these
82. Cucumber is the second widely cultivated cucurbit after
    (a) Long-melon (b) Water-melon
    (c) Snap-melon (d) None of these
83. The open pollinated cultivars of cucumber are in the sex form of
    (a) Dioecious (b) Monoecious
    (c) Both (d) None of these
84. Pusa Sanyog is the first F_1 hybrid of
    (a) Cucumber (b) Chilli
    (c) Okra (d) None of these
85. In India the first F_1 hybrid of cucumber Pusa Sanyog was developed in the year
    (a) 1969 (b) 1970
    (c) 1971 (d) 1972
86. 'Bronzing' resistant variety of cucumber is
    (a) Pusa Sanyog (b) Himangi
    (c) Both (d) None of these
87. Hybrid Pusa Sanyog is a cross between Japanese Gynoecious line and ....
    (a) Kheera-75 (b) Kheera-90
    (c) Green Long of Naples (d) None of these
88. In cucumber application of Ethrel in 150-200 ppm increases the
    (a) Number of female flowers (b) Fruit-set
    (c) Fruit yield (d) All the above
89. Application of 200-300 ppm Silver nitrate in gynoecious cucumber induces
    (a) Male flowers (b) Female flowers
    (c) Both (d) None of these
90. In cucumber, chemicals/plant regulators are applied at true-leaf stage of
    (a) 2 (b) 4
    (c) 6 (d) None of these
91. The cucurbit which contains high carotene is
    (a) Pumpkin (b) Cucumber
    (c) Both (d) None of these

92. *Cucurbita ficifolia* is known as
(a) Malabar gourd (b) Fig leaf gourd
(c) Both (d) None of these

93. Flowers of pumpkin are
(a) Unisexual (b) Bisexual
(c) Both (d) None of these

94. Flower colour of pumpkin is
(a) Lemon
(b) Lemon-yellow to deep orange
(c) White purple (d) None of these

95. Carotene Rich variety of Pumpkin is
(a) Arka Chandan (b) Pusa Vikas
(c) Both (d) None of these

96. Pumpkin variety suitable for cultivation in spring summer season in north India is
(a) Arka Chandan (b) Pusa Vikas
(c) Both (d) None of these

97. In pumpkin, high nitrogen application under high temperature conditions promotes
(a) Maleness (b) Femaleness
(c) Both (d) None of these

98. *Citrullus vulgaris* is the botanical name of
(a) Pumpkin (b) Water-melon
(c) Long-melon (d) None of these

99. Powdery mildew and red pumpkin beetle resistant variety of water-melon is
(a) Arka Manik (b) Pusa Bedana
(c) Both (d) None of these

100. Seedless variety of water melon is
(a) Arka Manik (b) Pusa Bedana
(c) Arka Jyoti (d) Sugar Baby

101. Pusa Bedana is an F_1 hybrid between tetra-2 and
(a) Sugar Baby (b) Madhu Melon
(c) Pusa Rasal (d) None of these

102. Average sweetness in water-melon is recommended at TSS % of
(a) 4-6 (b) 6-8
(c) 9-10 (d) None of these

103. *Cucumis melo* L. is the botanical name of
(a) Pumpkin (b) Water-melon
(c) Musk-melon (d) None of these

104. Average diameter of musk-melon fruit is
(a) 4-8 cm (b) 8-16 cm
(c) 16-32 cm (d) None of these

105. Anthesis in musk-melon takes place at
(a) 5.30-6.30 a.m. (b) 6.30-7.30 a.m.
(c) 7.30-8.30 a.m. (d) None of these

106. Suitable temperature for anthesis in musk-melon is
(a) 10-12°C (b) 15-16°C
(c) 22-29°C (d) None of these

107. Fruit setting in monoecious lines of *cucumis melo* is
(a) 10-12% (b) 12-20%
(c) 20-29% (d) 29-42%

108. The variety with excellent keeping quality of musk-melon is
(a) Arka Rajhans (b) Arka Jeet
(c) Pusa Madhu Ras (d) Durgapur Madhu

109. During fruit development the melons require temperature of (°C)
(a) 20-25 (b) 25-30
(c) 30-35 (d) 35-40

110. Optimum germination temperature in cucurbits is
(a) Above 25°C (b) Less 25°C
(c) Above 30°C (d) None of these

111. Muskmelon is a
(a) Climacteric fruit (b) Non-climacteric fruit
(c) Both (d) None of these

112. *Lagenaria siceraria* (Mol.) Standl is the botanical name of
(a) Bottle gourd (b) Round gourd
(c) Sponge gourd (d) None of these

113. Pusa Manjari is a round fruited F_1 hybrid of
(a) Bottle gourd (b) Round gourd
(c) Sponge gourd (d) None of these

114. Pusa Manjari is a round fruited F_1 hybrid of cross between Pusa Summer Prolific Round and
(a) Selection 10 (b) Selection 11
(c) Both (d) None of these

115. Fruits of Pusa Naveen variety of bottle gourd are
(a) Long (b) Cylindrical
(c) Round (d) None of these

116. High yielding variety of bottle gourd is
    (a) Pusa Summer Prolific Round
    (b) Pusa Summer Prolific Long
    (c) Both (d) None of these

117. 'Crook-neck' devoid variety of bottle gourd is
    (a) Pusa Naveen (b) Arka Bahar
    (c) Both (d) None of these

118. Variety of bottle gourd suitable for box packing and good keeping is
    (a) Pusa Naveen (b) Arka Bahar
    (c) Samrat (d) None of these

119. Early maturing variety of bottle gourd is
    (a) Pusa Hybrid-1 (b) Arka Bahar
    (c) Pusa Hybrid-3 (d) None of these

120. Short-day and humid climate promote the production in bottle gourd
    (a) Maleness (b) Femaleness
    (c) Both (d) None of these

121. Training of bottle gourd plants on bower system increases
    (a) Yield (b) Total fruit yield
    (c) High marketable yield (d) None of these

122. The cultivar of bottle gourd responding to bower system of training is
    (a) Arka Bahar (b) Samrat
    (c) Both (d) None of these

123. For getting higher number of female flowers in bottle gourd the spray of boron is recommended in a concentration of
    (a) 20 ppm (b) 10 ppm
    (c) 3 ppm (d) None of these

124. *Benincasa hispida* (Thumb) is the botanical name of
    (a) Ash gourd (b) Wax gourd
    (c) White gourd (d) All the above

125. Petha is also known as
    (a) Indian gourd
    (b) Chinese preserving melon
    (c) Both (d) None of these

126. Mudliar is an improved variety of
    (a) Ash gourd (b) Indian gourd
    (c) Musk-melon (d) None of these

127. The climate suitable for production of female flowers in ash gourd is
(a) Short-day (b) Low night temperature
(c) Humid climate (d) All the above

128. The seed rate (kg/ha) in ash gourd is
(a) 5-6 (b) 6-7
(c) 7-8 (d) None of these

129. Pusa Shakti is the improved variety of
(a) Ash gourd (b) Long melon
(c) Sponge gourd (d) None of these

130. *Luffa cylindrica* is the botanical name of
(a) Sponge gourd (b) Ash gourd
(c) Ridge gourd (d) None of these

131. The fruits of sponge gourd are almost cylindrical in shape with average length of
(a) 20-25 cm (b) above 50 cm
(c) above 60 cm (d) None of these

132. The flowers of sponge gourd open in early morning at
(a) 3-4 a.m. (b) 4-8 a.m.
(c) 8-10 a.m. (d) None of these

133. The Sponge gourd fruits are richer than ridge gourd in
(a) Protein (b) Carotene
(c) Protein and carotene (d) None of these

134. Sponge gourd contains a gelatinous compound called
(a) Luccim (b) Luffein
(c) Saponine (d) None of these

135. Fruits of sponge gourd are smooth and are called
(a) Ghia (b) Ghia tori
(c) Ribbed tori (d) None of these

136. Pusa Chikni and Pusa Supriya are the improved varieties of
(a) Sponge gourd (b) Ridge gourd
(c) Both (d) None of these

137. Kalyanpur Hari Chikni is the improved variety of
(a) Sponge gourd (b) Ridge gourd
(c) Both (d) None of these

138. *Luffa acutangula* (L.) Roxb is the botanical name of
(a) Sponge gourd (b) Ridge gourd
(c) Indian gourd (d) None of these

139. Ridge gourd is also known as
(a) Ribbed gourd (b) Angled lufta
(c) Both (d) None of these

140. The fruits of ridge gourd are club shaped and measure
(a) 8-10 cm (b) 20-50 cm
(c) 10-15 cm (d) None of these

141. Satputia is a cultivar of
(a) Sponge gourd (b) Ridge gourd
(c) Both (d) None of these

142. Satputia is mostly cultivated in
(a) Uttar Pradesh (b) Bihar
(c) Assam (d) None of these

143. Hermaphrodite cultivar of ridge gourd is
(a) Kalyanpur Hari Chikni (b) Satputia
(c) Both (d) None of these

144. Pusa Nasdar, Konkan Harita and Punjab Sadabahar are the improved varieties of
(a) Ridge gourd (b) Sponge gourd
(c) Both (d) None of these

145. Pumpkin as well as squashes do best at a soil pH of *(ARS, 2003)*
(a) 4.0-5.5 (b) 6.0-6.5
(c) 7.5-8.5 (d) None of these

146. Pumpkin is originated in *(ARS, 2003)*
(a) India (b) Africa
(c) America (d) None of these

147. The single seeded cucurbitaceous vegetable is *(ARS, 2003)*
(a) Chayote (b) Kakrol
(c) Kartoli (d) Ivy gourd

148. A cucurbitaceous crop considered to be perennial is *(ARS, 2003)*
(a) Parwal (b) Chow-chow
(c) Snake gourd (d) None of these

149. A triploid water-melon variety, Pusa Bedana was released in *(ARS, 2003)*
(a) 1970 (b) 1969
(c) 1972 (d) 1989

150. The seed rate (kg/ha) of round gourd is *(ARS, 2003)*
(a) 3-5 (b) 5-8
(c) 8-12 (d) 13-15

151. The root tubers of pointed gourd are enough to plant an area of one hectare is *(ARS, 2003)*
(a) 20,000 (b) 30,000
(c) 40,000 (d) 50,000

152. Variety of sponge gourd recommended by IARI is *(ARS, 2003)*
(a) Pusa Nasdar (b) Pusa Chikni
(c) Yard Long (d) None of these

153. Variety of winter squash recommended by IIHR, Bangalore is *(ARS, 2003)*
(a) Arka Chandan (b) Arka Suryamukhi
(c) Large Red (d) None of the above

154. For getting good fruit yield in parwal, planting ratio of female to male plants should be *(ARS, 2003)*
(a) 15-20 to 1 (b) 30-40 to 1
(c) 40-50 to 1 (d) None of these

155. Which of the following is the variety of water-melon? *(ARS, 2003)*
(a) Pusa Joyti (b) Arka Jyoti
(c) Arka Nishant (d) None of these

156. Possibility of producing seedless water-melon was first demonstrated in *(ARS, 2003)*
(a) U.S.A. (b) Japan
(c) Holland (d) None of these

157. First picking in bitter gourd can be done after *(ARS, 2003)*
(a) 50-60 days (b) 70-80 days
(c) 80-90 days (d) 90-100 days

158. Edible part of cucumber (pepo) is *(ARS, 2003)*
(a) Pericarp (b) Fleshy receptacle
(c) Mesocarp, endocarp and placentae
(d) Placentae

159. The most common cucurbit grown under glasshouses is *(ARS, 2003)*
(a) Bottle gourd (b) Watermelon
(c) Cucumber (d) None of these

160. Hermaphrodite cultivar of ridge gourd is *(ARS, 2003)*
(a) Pusa Nasdar (b) Satputia
(c) CO-1 (d) Jhinga Torai

161. Heterosis in bottle gourd was reported in a number of crosses by *(Ph.D. VGS, 2003)*
(a) I. D. Tyagi (1973) (b) V. S. Seshadri (1988)
(c) G. S. Gill (1980) (d) U. K. Verma (1985)

162. The most popular variety of water-melon is *(ARS, 2003)*
(a) Ashi Yamato (b) Shiper
(c) Sugar Baby (d) None of these

163. Kundru is propagated by *(ARS, 2003)*
(a) Seeds (b) Root cuttings
(c) Vine cuttings (d) None of these

164. An F_1 hybrid of cucumber recommended by IARI is *(ARS, 1992)*
(a) Poinsette (b) Pusa Sanyog
(c) Straight Eight (d) None of these

165. The most common sex expression in cucurbits is *(Ph.D. VGS, 2006)*
(a) Dioecism (b) Monoecism
(c) Andromonoecism (d) Androdioecism

166. Gynoecious lines are available in *(Ph.D. VGS, 2006)*
(a) Karela (b) Cucumber
(c) Watermelon (d) Musk-melon

167. Bottle gourd originated in *(Ph.D. VGS, 2006)*
(a) Africa (b) America
(c) India (d) China

178. Pumpkin originated in *(Ph.D. VGS, 2006)*
(a) Africa (b) America
(c) India (d) China

169. Seed rate (kg/ha) of water-melon is *(ARS (Pre), 2003)*
(a) 2-3 (b) 4-5
(c) 3-5 (d) 5-6

170. Seedlessness in watermelon is due to *(Ph.D. VGS, 2006)*
(a) Parthenocarpy (b) Triploidy
(c) Parthenogenesis (d) Embryo abortion

171. For local market harvesting of musk-melon should be done at *(ARS (Pre), 2003, ARS, 1999)*
(a) Full slip stage (b) Half slip stage
(c) Mature stage (d) Immature stage

172. Which part of the cucurbits is used as vegetables? *(ARS, 1992)*
(a) Succulent leaves (b) Succulent stem
(c) Fruits (d) None of these

173. The dioecious cucurbit is *(UP PCS (Pre), 2003)*
(a) Pumpkin (b) Pointed gourd
(c) Bottle gourd (d) Snake gourd

174. Sugar Baby is the variety of *(Ph.D. VGS, 2006)*
(a) Musk-melon (b) Pumpkin
(c) Water-melon (d) Bitter gourd

175. Chromosome number of bottle gourd (2n) is *(ARS, 1999)*
    (a) 22 (b) 24
    (c) 36 (d) 28

176. Optimum temperature for fruit setting in cucumber ranges between *(ARS, 1999)*
    (a) 55°F - 65°F (b) 70°F - 75°F
    (c) 40°F - 50°F (d) None of these

177. Cucumber is used for *(ARS, 1999)*
    (a) Fermenting (b) Pickling
    (c) Salad and Pickling (d) None of these

178. Ivy gourd is botanically called as *(ARS, 1999)*
    (a) *Cucumeropsis* (b) *Coccinia grandis*
    (c) *Trichosanthes dioica* (d) None of these

179. The diploid (2n) chromosome number of *Lagenaria siceraria* is *(ARS, 1999)*
    (a) 12 (b) 16
    (c) 22 (d) 24

180. The pointed gourd is classified as *(ARS, 1999)*
    (a) Monoecious (b) Adromonoecious
    (c) Gynomonoecious (d) Dioecious

181. The optimum seed rate kg/ha of cucumber is *(ARS, 1999)*
    (a) 1-1.5 (b) 2.5-3.5
    (c) 4.5-6.0 (d) 6.5-7.0

182. Triploid water-melons have *(ARS, 1999)*
    (a) Few viable seeds (b) More viable seeds
    (c) No viable seeds (d) Large seeds

183. The ratio of female and male plants in case of Parwal is *(ARS, 1999)*
    (a) 2 : 1 (b) 5 : 1
    (c) 6 : 1 (d) 10 : 1

184. The recommended N: P_2O_5:K_2O dose/ha in musk-melon is *(ARS, 1999)*
    (a) 60:30:30 (b) 75:30:30
    (c) 100:60:60 (d) 150:90:90

185. Fruit fly resistant variety of pumpkin is *(ARS, 1999)*
    (a) Arka Chandan (b) Arka Suryamukhi
    (c) Azad Pumpkin (d) None of these

186. Generally seedless water-melon is *(Ph.D. VGS, 2004)*
    (a) Diploid (b) Triploid
    (c) Tetraploid (d) Hexaploid

187. Pusa Bendana and Durgapur Kesar are the improved varieties of *(Ph.D. VGS, 2004)*
   (a) Water-melon (b) Musk-melon
   (c) Sweet potato (d) Beet

188. Sowing time of water-melon is *(Ph.D. VGS, 2005)*
   (a) August to October (b) February to April
   (c) November to January (d) May to July

189. Which crop is dioecious *(Ph.D. VGS, 2005)*
   (a) Bottle gourd (b) Sponge gourd
   (c) Pointed gourd (d) Bitter gourd

190. Pusa Bedana water-melon is a cross of *(UP PCS (Pre), 2006)*
   (a) Tetra-2 x Pusa Rasal (b) Pusa Rasal x Tetra-2
   (c) M.S.-1 x Hara Madhu (d) Hara Madhu x M.S.-1

191. Water-melon belongs to the species *(ARS, 2003)*
   (a) *sativus* (b) *vulgaris*
   (c) *lanatus* (d) None of these

192. Water-melon can be planted by *(ARS, 2003)*
   (a) Shallow pit method (b) Deep pit method
   (c) Ridge method (d) All of these

193. Water-melon is one of the few vegetables that grows well on acidic soil with pH of *(ARS, 2003)*
   (a) 4.0 (b) 5.0
   (c) 6.0 (d) None of these

194. Chayote is generally grown by planting *(ARS, 2003)*
   (a) Seeds (b) Tubers
   (c) Vine cuttings (d) Entire

195. Fruits of wild forms of bottle gourd are bitter in taste due to *(ARS, 2003)*
   (a) Solanin (b) Tomatine
   (c) Cucurbitacin (d) None of these

196. Pumpkin varieties mature in *(ARS, 2003)*
   (a) 70-80 days (b) 80-90 days
   (c) 110-120 days (d) None of these

197. Under good management once planted pointed gourd can give economical returns for at least *(ARS, 2003)*
   (a) 2 years (b) 3 years
   (c) 4 years (d) 1 year

ANSWERS

1. (a)	2. (b)	3. (c)	4. (c)	5. (c)
6. (a)	7. (a)	8. (b)	9. (b)	10. (b)

11. (a)	12. (d)	13. (d)	14. (b)	15. (c)
16. (b)	17. (c)	18. (c)	19. (b)	20. (c)
21. (a)	22. (a)	23. (a)	24. (b)	25. (c)
26. (a)	27. (a)	28. (c)	29. (b)	30. (c)
31. (c)	32. (b)	33. (a)	34. (b)	35. (a)
36. (b)	37. (a)	38. (d)	39. (c)	40. (a)
41. (d)	42. (c)	43. (c)	44. (a)	45. (c)
46. (a)	47. (d)	48. (a)	49. (a)	50. (b)
51. (b)	52. (c)	53. (c)	54. (b)	55. (c)
56. (b)	57. (a)	58. (c)	59. (a)	60. (c)
61. (a)	62. (c)	63. (a)	64. (d)	65. (d)
66. (b)	67. (b)	68. (b)	69. (c)	70. (b)
71. (b)	72. (b)	73. (a)	74. (a)	75. (a)
76. (b)	77. (c)	78. (b)	79. (a)	80. (b)
81. (a)	82. (b)	83. (b)	84. (a)	85. (c)
86. (b)	87. (c)	88. (d)	89. (a)	90. (a)
91. (a)	92. (c)	93. (a)	94. (b)	95. (a)
96. (b)	97. (a)	98. (b)	99. (a)	100. (b)
101. (c)	102. (c)	103. (c)	104. (b)	105. (a)
106. (c)	107. (d)	108. (a)	109. (d)	110. (a)
111. (a)	112. (a)	113. (a)	114. (b)	115. (b)
116. (a)	117. (b)	118. (c)	119. (c)	120. (b)
121. (a)	122. (b)	123. (c)	124. (d)	125. (b)
126. (a)	127. (d)	128. (a)	129. (a)	130. (a)
131. (a)	132. (b)	133. (c)	134. (b)	135. (b)
136. (a)	137. (a)	138. (b)	139. (c)	140. (c)
141. (b)	142. (b)	143. (b)	144. (a)	145. (b)
146. (c)	147. (a)	148. (b)	149. (c)	150. (b)
151. (d)	152. (b)	153. (b)	154. (a)	155. (b)
156. (b)	157. (a)	158. (c)	159. (c)	160. (b)
161. (a)	162. (c)	163. (c)	164. (b)	165. (b)
166. (c)	167. (a)	168. (b)	169. (b)	170. (d)
171. (a)	172. (c)	173. (b)	174. (c)	175. (a)
176. (a)	177. (c)	178. (b)	179. (c)	180. (d)
181. (b)	182. (c)	183. (d)	184. (c)	185. (b)
186. (b)	187. (a)	188. (b)	189. (c)	190. (a)
191. (b)	192. (d)	193. (b)	194. (d)	195. (c)
196. (c)	197. (c)			

CHAPTER 6

LEGUMINOUS VEGETABLES

Peas, cowpea, cluster bean, french bean and hyacinth bean are important leguminous vegetables which belong to the family leguminosae.

6.1 Pea

- *Pisum elatius*, a wild species, is considered as the ancestor of pea (*Pisum sativum*).
- Peas are diploid with a chromosome number of 2n=14.
- Pea is a herbaceous winter annual.
- Arka Ajit is resistant to powdery mildew and rust.
- UN 53 (6) is a snap pea (whole pod edible) line developed by IIHR, Bangalore.
- Arkel is the most popular exotic pea introduced from England.
- Bonneville is a popular variety in India introduced from USA.
- Flowering, fruit set and grain filling periods are the critical stages for irrigation in peas.

6.2 Cowpea

- Cowpea is used as fodder, green manure and vegetable crop.
- Cowpea belongs to the family *Leguminosae,* sub-family *Fabaceae*.
- Pusa Dofasli, Arka Garima and Arka Suman are photo-insensitive varieties, suitable for both summer and rainy seasons.
- Cowpea pods are rich in vitamin B.
- Cowpea is grown as a catch crop.
- Flowering and pod development periods are the critical stages for water management in cowpea.

6.3 French bean

- The French bean is also known as kidney bean, haricot bean, snap bean, navy bean, and is one of the most important leguminous vegetables.

- There are three types of French bean: Bush type with short internodes, semi-pole type with longer internodes than those in bush type and pole type having longer internodes than that of semi-pole-type.
- Pusa Parvati is resistant to mosaic and powdery mildew.
- Kentucky Wonder and Contender are introduced varieties from USA.
- Through pure-line selection, Arka Komal, Pant Anupama and IIHR 220 have been developed.
- Variety IIHR 220 is resistant to rust.
- Through mutation breeding, Pusa Parvati has been developed and released from IARI, Katrain.
- French bean is a shallow rooted crop.
- Staking is an important operation for pole beans.
- Pusa Early Prolific is a pole-type bean, suitable for autumn and spring seasons.
- Arka Jay and Arka Vijay are bush type beans, photo-insensitive, tolerant to heat and drought.
- Indian bean is a cool season and drought tolerant crop.
- Lima bean is also known as double bean or butter bean.
- Lima bean, though of high nutritional value, is not commonly grown in India.

6.4 Cluster bean and other beans

- Cluster bean is grown for green pods used as vegetable and dry seeds.
- Pusa Sadabahar and Pusa Navbahar are improved varieties of cluster bean.
- Variety, Pusa Sadabahar is non-branching and suitable for summer and rainy season.
- Cluster bean is a hardy and drought tolerant crop.
- Cluster bean is grown as a sole crop, mixed crop and a border crop around a main crop.
- Dolichos bean or Indian bean or Hyacinth is grown for its whole pods.
- Indian bean is a herbaceous perennial, but usually grown from seeds as an annual.
- The broad bean is also known as faba bean or horse bean.
- Broad bean is widely cultivated in Latin America.
- Broad bean tolerates salinity.
- Broad bean is mainly grown at higher elevation.
- Winged bean is also known as Four-angled bean or Goa bean.

- Winged bean is a herbaceous, perennial, tropical legume of high value.
- Winged bean is photosensitive and requires short days for normal flower induction.
- Winged bean is free from pests and diseases. Recently minor diseases like false rust and leaf spot and pest like aphids were observed.

The objective type questions on leguminous vegetables are given below :

1. *Pisum sativum* L. sp. *hortense* is the botanical name of
   (a) Field pea (b) Garden pea
   (c) Both (d) None of these
2. *Pisum sativum* L. sp. *arvense* is the botanical name of
   (a) Field pea (b) Garden pea
   (c) Both (d) None of these
3. Flower colour of field pea is
   (a) White (b) Pink
   (c) Both (d) None of these
4. Flower colour of garden pea is
   (a) White (b) Red
   (c) Both (d) None of these
5. Garden pea is also known as
   (a) Horticultural pea (b) Sweet pea
   (c) Both (d) None of these
6. Diploid chromosome number of pea is
   (a) 7 (b) 14
   (c) Both (d) 24
7. Powdery mildew resistant variety of pea is
   (a) Arkel (b) Arka Ajit
   (c) Azad P-1 (d) Azad P-2
8. White fly resistant variety of pea is
   (a) Azad P-3 (b) Azad P-5
   (c) Ooty-1 (d) None of these
9. If pea is sown earlier it suffers from the disease
   (a) Wilt (b) Mildew
   (c) Both (d) None of these
10. If pea is sown late it suffers from the disease
    (a) Wilt (b) Powdery mildew
    (c) Both (d) None of these
11. Azad P-3 is a variety of pea belonging to maturity group
    (a) Early (b) Mid
    (c) Late (d) All the above

12. Pea varieties developed through introduction are
    (a) Bonneville (b) Arkel
    (c) Both (d) None of these
13. Whole pod edible variety of pea is
    (a) Madhu (b) Sylvia
    (c) Both (d) None of these
14. Extra early variety of pea is
    (a) Azad P-3 (b) Harbhajan
    (c) Arkel (d) None of these
15. In the processing industry the maturity of pea is tested with the help of
    (a) Refectometer (b) Tenderometer
    (c) Both (d) None of these
16. Pant Uphar is a variety of
    (a) French bean (b) Pea
    (c) Cowpea (d) None of these
17. Critical stage for irrigation in pea is
    (a) Flowering (b) Fruit setting
    (c) Grain filling (d) All the above
18. Origin of vegetable pea is
    (a) Africa (b) India
    (c) Central Asia (d) None of these
19. *Vigna unguiculata* is the botanical name of
    (a) French bean (b) Cowpea
    (c) Pea (d) None of these
20. Seed rate (kg/ha) of cowpea is
    (a) 10-12 (b) 15-20
    (c) 20-25 (d) None of these
21. Cowpea is a rich source of
    (a) Vit-A (b) Vit-B
    (c) Vit-C (d) None of these
22. Cowpea is used for
    (a) Fodder (b) Green manure
    (c) Vegetable (d) All the above
23. Cowpea belongs to the family
    (a) *Leguminosae* (b) *Apiaceae*
    (c) *Fabaceae* (d) None of these
24. Sub-family of cowpea is
    (a) *Leguminosae* (b) *Apiaceae*
    (c) *Fabaceae* (d) None of these

25. Photoinsensitive variety of cowpea is
    (a) Pusa Dofasli (b) Arka Garima
    (c) Arka Suman (d) All the above
26. Cowpea is a
    (a) Cash crop (b) Catch crop
    (c) Both (d) None of these
27. Critical stages of irrigation in cowpea is
    (a) Flowering (b) Pod development
    (c) Both (d) None of these
28. *Phaseolus vulgaris* L.is the botanical name of
    (a) Cowpea (b) French bean
    (c) Cluster bean (d) None of these
29. French bean is also known as
    (a) Haricot bean (b) Kidney bean
    (c) Snap bean/navy bean (d) All the above
30. French bean progenitor is
    (a) *P. Vulgaris* (b) *P. aborigineus*
    (c) *P. coccineus* (d) None of these
31. In respect of pollination, French bean is
    (a) Self pollinated (b) Cross pollinated
    (c) Both (d) None of these
32. *Phaseolus coccineus* species of French bean is
    (a) Self pollinated (b) Cross pollinated
    (c) Both (d) None of these
33. Mosaic and powdery mildew resistant variety of Rajma is
    (a) Arka Komal (b) Pusa Parvati
    (c) Pant Anupama (d) None of these
34. Introduced variety of French bean is
    (a) Kentucky Wonder (b) Contender
    (c) Both (d) None of these
35. Pure-line selected variety of French bean is
    (a) Arka Komal (b) Pant Anupama
    (c) IIHR-220 (d) All the above
36. Rust resistant variety of Rajma is
    (a) IIHR-220 (b) IIHR-990
    (c) Both (d) None of these
37. Mutant variety of french bean is
    (a) IIHR-990 (b) Pusa Parvati
    (c) Both (d) None of these

38. Pusa Parvati variety of french bean was developed from
    (a) IARI (b) Katrain
    (c) IIPR (a) None of these
39. French bean is a
    (a) Tap rooted crop (b) Shallow rooted crop
    (c) Both (d) None of these
40. Critical stage for irrigation in French bean is at
    (a) Flowering (b) Pod development
    (c) Both (d) None of these
41. *Cyamopsis tetragonoloba* (L.) Taub. is the botanical name of
    (a) Cowpea (b) Cluster bean
    (c) French bean (d) None of these
42. Cluster bean is also known as
    (a) Cowpea (b) Guar
    (c) Both (d) None of these
43. Variety of Cluster bean suitable for summer and rainy seasons is
    (a) Pusa Sadabahar (b) Pusa Navbahar
    (c) Pusa Mausami (d) All the above
44. Cluster bean is grown as a
    (a) Sole crop (b) Mixed crop
    (c) Border crop (d) All the above
45. Critical stage for irrigation in cluster bean is at
    (a) Flowering (b) Pod development
    (c) Both (d) None of these
46. *Dolichos* bean is also known as
    (a) Indian bean (b) Hyacinth bean
    (c) Both (d) None of these
47. Pole type variety of *Dolichos* bean is
    (a) Arka Jay (b) Pusa Early Prolific
    (c) Both (d) None of these
48. Indian bean variety suitable for both autumn and spring seasons is
    (a) Arka Jay (b) Rajni
    (c) Pusa Early Prolific (d) None of these
49. The variety of Indian bean tolerant to heat and drought is
    (a) Arka Jay (b) Arka Vijay
    (c) Both (d) None of these
50. Dual purpose type variety of Indian bean is
    (a) Arka Vijay (b) Konkan Bhusan
    (c) Both (d) None of these

51. Indian bean is grown during
    (a) Summer (b) Winter
    (c) Both (d) None of these
52. Bush type varieties of Indian bean are ready for harvest in
    (a) One month (b) Two months
    (c) Both (d) None of these
53. Pole type varieties of Indian bean are ready for harvest in
    (a) One month (b) Two months
    (c) Three months (d) None of these
54. Lima bean is also known as
    (a) Double bean (b) Butter bean
    (c) Both (d) Single bean
55. High nutritional value vegetable is
    (a) Indian bean (b) Lima bean
    (c) Winged bean (d) None of these
56. Pole type varietiy of Lima bean is
    (a) King of Garden (b) Challenger
    (c) Florida Butter (d) All the above
57. Semi-pole type variety of Lima bean is
    (a) Wilbur (b) Hopi
    (c) Both (d) None of these
58. Baby Potato and Baby Ford Hook are the improved varieties of
    (a) Indian bean (b) Lima bean
    (c) Broad bean (d) None of these
59. *Vicia faba* L. is the botanical name of
    (a) Lima bean (b) Broad bean
    (c) Winged bean (d) None of these
60. Green Long Pod, Imperial Green Long Pod and Red Epicure are the improved varieties of
    (a) Lima bean (b) Broad bean
    (c) Winged bean (d) None of these
61. Bean grown as both autumn and winter crops is
    (a) Lima bean (b) Broad bean
    (c) Winged bean (d) None of these
62. *Phaseolus tetragonolobus* (L.)DC. is the botanical name of
    (a) Lima bean (b) Broad bean
    (c) Winged bean (d) None of these
63. Chromosome number of Winged bean is
    (a) 2n = 10 (b) 2n = 12
    (c) 2n = 16 (d) 2n = 18

64. Winged bean is also known as
    (a) Four-angled bean (b) Goa bean
    (c) Both (d) None of these

65. Winged bean is
    (a) Herbaceous (b) Perennial
    (c) Tropical legume (d) All the above

66. Giant chillies are the richest source of thiamin followed by
    (a) Potato (b) Pea
    (c) Both (d) None of these

67. French bean is a native of *(ARS 2003)*
    (a) Asia (b) Mediterranean region
    (c) South and Central America (d) None of the above

68. French bean variety resistant to anthracnose disease is *(ARS, 2003)*
    (a) Contender (b) Kentucky Wonder
    (c) Tweed Wonder (d) None of these

69. Fresh unshelled peas can be stored well for two weeks at *(ARS, 2003)*
    (a) 5°C and 85% R.H. (b) 0°C and 90-95% R.H.
    (c) 15°C and 60-70% R.H. (d) None of the above

70. The variety of Lima bean which needs support is *(ARS, 2003)*
    (a) Pole type (b) Semi-pole type
    (c) Bush type (d) All of these

71. A guar meal contains protein *(ARS, 2003)*
    (a) 11.3% (b) 22.2%
    (c) 33.3% (d) 44.4%

72. A Pea variety which gives 50% produce during first picking is *(ARS, 2003)*
    (a) Arkel (b) Meteor
    (c) Mattar Ageta-6 (d) None of these

73. Cluster bean contains mucilaginous substance called *(ARS, 2003)*
    (a) Allantoin (b) Allantonic acid
    (c) Mannogalacton (d) Glycoprotein

74. Cluster bean is sown at a spacing of *(ARS, 2003)*
    (a) 75 x 60 cm (b) 60 x 45 cm
    (c) 45 x 20 cm (d) None of these

75. Cluster bean is *(ARS, 2003)*
    (a) Self pollinated crop (b) Cross pollinated crop
    (c) Often cross pollinated crop
    (d) None of the above
76. Cluster bean which produces an amide is *(ARS, 2003)*
    (a) Aspargine (b) Allantion
    (c) Glycoprotein (d) Allantonic acid
77. Contender is a variety of *(ARS, 2003)*
    (a) French bean (b) Broad bean
    (c) Lima bean (d) None of these
78. Cowpea is highly susceptible to the deficiency of *(ARS, 2003)*
    (a) Zn (b) B
    (c) Mo (d) Mn
79. Cowpea is native of *(ARS, 2003)*
    (a) India (b) China
    (c) Central Africa (d) None of these
80. Cowpea variety which is highly resistant to mosaic virus is *(ARS, 2003)*
    (a) Pusa Dofasli (b) Cowpea 263
    (c) Pusa Komal (d) None of these
81. Vitamin C content in French bean is enhanced by the application of minor element *(ARS, 2003)*
    (a) Zinc (b) Iron
    (c) Copper (d) None of these
82. Challenger is a variety of *(ARS, 2003)*
    (a) Indian bean (b) Lima bean
    (c) Cluster bean (d) Hyacinth bean
83. Cluster bean belongs to the genus *(ARS 2003, Ph.D. VGS, 2006)*
    (a) *Phaseolus* (b) *Dolichos*
    (c) *Cyamopsis* (d) None of these
84. Pusa Barsati is the variety of *(ARS, 2003)*
    (a) Cowpea (b) Pea
    (c) French bean (d) Cluster bean
85. Depending upon the soil type and weather, pea crop requires total irrigations numbering *(ARS, 2003)*
    (a) 3-4 (b) 8-10
    (c) 15-16 (d) None of these
86. The most favorable pH for pea is *(JRF, 2005)*
    (a) 6.0-7.5 (b) Below 5.5
    (c) 7.5-8.5 (d) None of these

87. Growth habit of Henderson Bush variety of Lima bean is *(ARS, 2003)*
   (a) Pole type (b) Semi-pole type
   (c) Bush type (d) None of these
88. High quality of pea seed should have a minimum germination percentage of *(ARS, 2003)*
   (a) 75 (b) 85
   (c) 95 (d) 65
89. Highest pod yield in French bean is obtained at a soil pH of *(ARS, 2003)*
   (a) 5.3-6.0 (b) 6.0-7.0
   (c) 7.5-8.5 (d) None of these
90. Hyacinth bean or sem originated in *(ARS, 2003)*
   (a) America (b) India
   (c) China (d) None of these
91. The most serious viral disease of bean is *(ARS, 2003)*
   (a) Common bean mosaic (b) Yellow bean mosaic
   (c) Curly top (d) None of the above
92. Pea is very rich in *(ARS, 1992)*
   (a) Carbohydrates (b) Protein
   (c) Fat (d) Acid
93. The nitrogen fixation from air through symbiosis is brought about by the bacteria *(ARS, 1992)*
   (a) *Rhizobium japonicum* (b) *Rhizobium maeliloti*
   (c) *Rhizobium leguminosarum*
   (d) *Rhizobium species*
94. The bean used for making dalmoth is *(ARS, 1992)*
   (a) Cluster bean (b) Hyacinth bean
   (c) French bean (d) All of the above
95. Cowpea is native of *(ARS, 1992)*
   (a) India (b) Central Africa
   (c) South America (d) China
96. The most favourable pH for pea is *(Ph.D. VGS, 2006)*
   (a) 6.0 to 7.5 (b) 7.5 to 8.0
   (c) 8.0 to 8.5 (d) below 5.0
97. The pea variety which is not early is *(RAS (Pre), 2003)*
   (a) Arkel (b) Meteor
   (c) Early giant (d) Bonneville
98. In garden pea main season variety suitable for dehydration is *(Ph.D. VGS, 2006)*
   (a) Punjab 87 (b) Punjab 88
   (c) Bonneville (d) Arkel

99. The family of peas, beans and cowpea is *(Ph.D. VGS, 2006)*
(a) *Leguminosae* (b) *Solanaceae*
(c) *Liliaceae* (d) *Malvaceae*

100. The peas seeds become smooth after drying in pea variety *(Ph.D. VGS, 2006)*
(a) Arkel (b) Bonneville
(c) Rachna (d) Asauji

101. Leafless variety of pea is *(UP PCS (Pre), 1996, 2002)*
(a) Arkel (b) Azad Pea-1
(c) Aparna (d) L-116

102. Seed rate of peas (kg/ha) is *(Ph.D. VGS, 2006)*
(a) 150 (b) 140
(c) 120 (d) 100

103. The botanical name of Rajma (French bean) is *(UP PCS (Pre), 1998)*
(a) *Phaseolus vulgaris* (b) *Glycine max*
(c) *Lens esculenta* (d) *Cicer arietinum*

104. Chromosome number of fenugreek is *(Ph.D. VGS, 2006)*
(a) 2n = 16 (b) 2n = 26
(c) 2n = 36 (d) 2n = 4

105. Powdery mildew in pea is caused by *(UP PCS (Pre), 2005)*
(a) *Uromyces fabae* (b) *Colletotrichum pise*
(c) *Erysiphe polygoni* (d) *Peronospora pise*

106. Powdery mildew resistant variety of pea is *(UP PCS (Pre), 2004, 2005)*
(a) VL-1 (b) T-163
(c) Rachna (d) AP-3

107. The nitrogen requirement of French bean (kg/ha) is *(ARS, 1999)*
(a) 30 (b) 70
(c) 100 (d) 120

108. Pusa Dofasli is the variety of *(ARS, 1999)*
(a) Cucumber (b) Cowpea
(c) French bean (d) None of these

109. The bean used for extraction of gum is *(ARS, 1999)*
(a) Hyacinth bean (b) Cluster bean
(c) Yard long bean (d) None of these

110. The early variety of vegetable pea is *(ARS, 1999)*
(a) Arkel (b) NP-29
(c) Rachna (d) Bonneville

111. Seed rate of pea in (kg/ha) is *(ARS, 1999)*
(a) 100 (b) 120
(c) 140 (d) 150

112. Broad bean is a *(ARS, 1999)*
(a) Cool season crop (b) Warm season crop
(c) Temperate crop (d) None of these

113. Lima beans are *(ARS, 1999)*
(a) Highly self pollinated crops (b) Cross pollinated crops
(c) Often cross pollinated crops (d) None of these

114. Whole part edible bean crop is *(ARS, 1999)*
(a) Cowpea (b) Faba bean
(c) French bean (d) Winged bean

115. Arkel, Early Badger, Bonneville and Early December are the improved varieties of *(Ph.D. VGS, 2003)*
(a) Lentil (b) Pea
(c) Sweet potato (d) None of these

116. Pusa Navbahar and Pusa Mausmi are the improved varieties of
(a) Cowpea (b) Guar
(c) Oat (d) None of these

ANSWERS

1. (b)	2. (a)	3. (b)	4. (a)	5. (c)
6. (b)	7. (b)	8. (c)	9. (a)	10. (b)
11. (a)	12. (c)	13. (b)	14. (b)	15. (b)
16. (b)	17. (d)	18. (c)	19. (b)	20. (c)
21. (b)	22. (d)	23. (a)	24. (c)	25. (d)
26. (b)	27. (c)	28. (b)	29. (d)	30. (b)
31. (a)	32. (b)	33. (b)	34. (c)	35. (d)
36. (a)	37. (b)	38. (b)	39. (b)	40. (c)
41. (b)	42. (b)	43. (a)	44. (d)	45. (c)
46. (c)	47. (b)	48. (c)	49. (c)	50. (c)
51. (b)	52. (b)	53. (c)	54. (c)	55. (b)
56. (d)	57. (c)	58. (b)	59. (b)	60. (b)
61. (b)	62. (c)	63. (d)	64. (c)	65. (d)
66. (b)	67. (c)	68. (c)	69. (b)	70. (a)
71. (c)	72. (c)	73. (c)	74. (c)	75. (a)
76. (a)	77. (a)	78. (a)	79. (c)	80. (d)

81. (a) 82. (b) 83. (c) 84. (a) 85. (a)
86. (a) 87. (c) 88. (a) 89. (a) 90. (b)
91. (a) 92. (b) 93. (d) 94. (b) 95. (b)
96. (a) 97. (d) 98. (a) 99. (a) 100. (c)
101. (c) 102. (c) 103. (a) 104. (a) 105. (c)
106. (c) 107. (a) 108. (b) 109. (b) 110. (a)
111. (b) 112. (a) 113. (c) 114. (d) 115. (b)
116. (b)

CHAPTER 7

BULB CROPS

Onion, garlic and leek are the bulb crops belonging to the family *Amaryllidaceae.*

7.1 Onion

- The inflorescence in onion is an umbel.
- *Allium* belongs to class *Monocotyledonae*.
- The flowers of onion are white or bluish.
- Anther dehiscence usually occurs between 5 a.m. and 9 a.m.
- Onion is highly cross-pollinated crop due to protandry.
- The onion is pollinated chiefly by honey bees. Blow flies are excellent pollinators, as they go from flower to flower.
- In onion, male-sterile lines have been isolated in Pusa Red at IARI.
- Pusa Red is very good in storage and for wider adaptability.
- Arka Kalyan, Agrifound Dark Red, Baswant 780, N-53 and N2-4-1 are suitable under *kharif* season.
- Arka Bindu and Agrifound Rose are suitable for export, particularly to Malaysia and Singapore.
- Punjab 48 and Udaipur 102 are white skinned varieties of onion.
- Early Grano and Brown Spanish are yellow skinned varieties of onion.
- Agrifound Dark Red is multiplier onion.
- Long-day varieties of onion do not form bulb under short-day conditions, whereas short-day varieties, if planted under long-day conditions, will develop early bulbs.
- The usual method of propagation of common onion is by seeds.
- Onion is usually grown in winter season (*rabi*) in northern India, but it is grown both in *rabi* and rainy seasons (*kharif*) in Maharashtra, Gujarat, Andhra Pradesh, Karnataka and Tamil Nadu.
- The most favorable temperature for development of purple blotch (*Alternaria porri*) disease of onion is 28-30°C with 80-90% relative humidity.

- The varieties Early Grano, Pusa White Flat and Pusa White Round are recommended for growing as green onion.
- The average yield of multiplier onion is 15-18 tonnes/ha.
- For the *kharif* onion, spraying of maleic hydrazide 2,500 ppm at 75-90 days after transplanting checks the sprouting.
- Pusa White Flat, Pusa White Round, Punjab 48(S-4-8), N257-9-1 and Udaipur 102 have been identified for dehydration purposes.

7.2 Garlic

- China ranks first in area and production of garlic, whereas India ranks second in area and third in production.
- Garlic is used as a spice or condiment throughout India.
- Allicin is the anti-bacterial and has the typical odour of fresh garlic.
- Diallyl disulphide is said to possess true garlic odour.
- Garlic is a sexually sterile diploid.
- Yamuna Safed is recommended for cultivation all over India.
- Garlic bulbs are produced only in one season, i.e. winter (rabi).

7.3 Leek

- Leek is a non-bulb forming member of the onion family and is grown for its blanched stem and leaves.
- Leek belongs to the genus *Allium* and species *porrum.*
- The most common varieties of leek are London Flag and American Flag.

The important objective type questions on bulb crops are given below :

1. In area and production of onion in the world India stands
   (a) I (b) II
   (c) III (d) None of these
2. Leading onion producing state in India is
   (a) Maharashtra (b) Karnataka
   (c) Uttar Pradesh (d) None of these
3. State having highest productivity of onion is
   (a) Maharashtra (b) Karnataka
   (c) Gujarat (d) Punjab

4. Leading onion exporter country of the world is
   (a) India (b) China
   (c) Netherland (d) Bangladesh
5. Onion and garlic belong to the family
   (a) *Amaryllidaceae* (b) *Chenopodiaceae*
   (c) *Umbelliferae* (d) *Gulanaceae*
6. Chromosome number of onion and garlic is
   (a) 2n = 8 (b) 2n = 16
   (c) 2n = 24 (d) 2n = 48
7. Inflorescence of onion is
   (a) Corymb (b) Umbel
   (c) Cymose (d) Racemose
8. Allium belongs to class
   (a) *Dicotyledonae* (b) *Monocotyledonae*
   (c) Both (d) None of these
9. Flower colour of onion is
   (a) Yellow (b) Purple
   (c) White-bluish (d) Pink
10. Anthesis in onion takes place at
    (a) 5 a.m. (b) 9 a.m.
    (c) 5-9 a.m. (d) 9-12 a.m.
11. Onion is highly cross pollinated crop due to
    (a) Protogyny (b) Protandry
    (c) Both (d) None of these
12. Pollinater of onion is
    (a) Blow flies (b) Honey bees
    (c) Aphids (d) None of these
13. In onion male-sterile lines have been isolated at IARI in variety
    (a) Pusa Red (b) Kalyanpur Red Round
    (c) N-53 (d) Early Grano
14. The variety of onion which contains wider adaptability and good storage life is
    (a) Pusa Red (b) N-53
    (c) Baswant-780 (d) Kalyanpur Red Round
15. The onion varieties suitable for *kharif* season are
    (a) Arka Kalyan (b) Agrifound Dark Red
    (c) Baswant-780, N-53, N2-4-1
    (d) All the above
16. Onion variety suitable for export is
    (a) Arka Bindu (b) Agrifound Rose
    (c) Both (d) None of these

17. White coloured variety of onion is
    (a) Punjab 48 (b) Udaipur 102
    (c) Both (d) None of these
18. Yellow coloured variety of onion is
    (a) Early Grano (b) Brown Spanish
    (c) Both (d) None of these
19. Long-day type variety of onion is
    (a) Brown Spanish (b) Early Grano
    (c) N-53 (d) None of these
20. Multiplier onion variety is
    (a) Agrifound Light Red (b) Agrifound Dark Red
    (c) N-53 (d) None of these
21. Onion is propagated through
    (a) Seeds (b) Umbel
    (c) Roots (d) None of these
22. In Karnataka small pickling onion is planted .... time(s) in a year.
    (a) one (b) Two
    (c) Three (d) None of these
23. Optimum bulb size for obtaining good quality onion is
    (a) 1 - 1.5 cm (b) 1.5 - 2.0 cm
    (c) 2.0 - 2.5 cm (d) None of these
24. Pungency in onion is due to
    (a) Allyl propyl sulphide (b) Allyl propyl disulphide
    (c) Both (d) None of these
25. The colour of the outer skin of onion is due to
    (a) Aceteen (b) Quercetin
    (c) Both (d) None of these
26. Best soil pH for onion cultivation is
    (a) 4-5 (b) 5.8-6.5
    (c) 6.5-7.5 (d) None of these
27. Seed rate (g/ha) of onion is
    (a) 200-300 (b) 300-400
    (c) 400-500 (d) 500-600
28. Most favourable temperature for purple blotch disease of onion is
    (a) 10-15°C (b) 15-20°C
    (c) 28-30°C (d) > 30°C
29. Purple blotch is a common disease of
    (a) Garlic (b) Onion
    (c) Both (d) None of these

30. The variety recommended for green onion is
    (a) Early Grano (b) Pusa White Flat
    (c) Pusa White Round (d) All the above
31. Average yield (t/ha) of multiplier onion is
    (a) 15-18 (b) 20-25
    (c) 30-40 (d) None of these
32. Average yield (t/ha) of *kharif* onion is
    (a) 15-18 (b) 20-25
    (c) 30-40 (d) None of these
33. Cytoplasmic genetic male sterility was first reported in
    (a) Garlic (b) Onion
    (c) Peas (d) None of these
34. Anthesis starts in *Alliun cepa* at
    (a) 7 a.m. (b) 9 a.m.
    (c) 10 a.m. (d) None of these
35. *Allium sativum* L. is the botanical name of
    (a) Onion (b) Garlic
    (c) Leek (d) None of these
36. In area and production of garlic the country which ranks first is
    (a) India (b) China
    (c) Africa (d) None of these
37. India's rank in area and production of garlic is
    (a) First, Second (b) Second, Third
    (c) Second, Forth (d) None of these
38. Per capita/year availability of garlic in India is about
    (a) 1.00 kg (b) 0.750 kg
    (c) 0.34 kg (d) None of these
39. Leading garlic producing state is
    (a) Maharashtra (b) M.P.
    (c) Gujarat (d) None of these
40. The state having highest area under garlic cultivation is
    (a) Madhya Pradesh (b) Gujarat
    (c) Uttar Pradesh (d) None of these
41. The state having highest productivity of garlic is
    (a) Punjab (b) Haryana
    (c) Madhya Pradesh (d) Maharashtra
42. A Colourless, odourless and water soluble amino acid found in garlic is
    (a) Allin (b) Allicin
    (c) Both (d) None of these

43. Anti-bacterial substance found in garlic is
    (a) Allin (b) Allicin
    (c) Both (d) None of these
44. Garlic odour is due to
    (a) Diallyl disulphide (b) Allyl propyl disulphide
    (c) Both (d) None of these
45. Yamuna Safed is the improved cultivar of
    (a) Onion (b) Garlic
    (c) Leek (d) None of these
46. Export variety of garlic is
    (a) Yamuna safed (b) G-282
    (c) Both (d) None of these
47. Long-day variety of garlic is
    (a) Agrifound Parvati (b) Yamuna Safed
    (c) G-282 (d) None of these
48. In India garlic varieties are of .... type.
    (a) Long day (b) Short day
    (c) Both (d) None of these
49. Garlic is propagated through
    (a) Bulbs (b) Bulblets
    (c) Cloves (d) None of these
50. Garlic is produced in .... season.
    (a) One (b) Two
    (c) Three (d) None of these
51. Garlic is a
    (a) Vegetable (b) Clove
    (c) Condiment or spice (d) None of these
52. *Allium porrum* is the botanical name of
    (a) Garlic (b) Onion
    (c) Leek (d) None of these
53. Family of *Allium porrum* is
    (a) *Amaryllidaceae* (b) *Anacardiaceae*
    (c) *Leguminasae* (d) All the above
54. Non-bulb forming member of the onion family is
    (a) Garlic (b) Onion
    (c) Asparagus (d) Leek
55. Edible part of leek is
    (a) Stem (b) Leaves
    (c) Both (d) None of these
56. London Flag and American Flag are the improved varieties of
    (a) Garlic (b) Onion
    (c) Leek (d) None of these

57. Leek is
(a) Perennial (b) Biennial
(c) Annual (d) None of these

58. For getting high seed yield, bulbs of onion are planted at a spacing of *(ARS, 2003)*
(a) 60 x 45 cm (b) 45 x 45 cm
(c) 45 x 30 cm (d) None of these

59. For getting higher bulb yield, grow onion at a spacing of *(ARS, 2003)*
(a) 30 x 45 cm (b) 30 x 7.5 cm
(c) 15 x 7.5 cm (d) None of these

60. The loss due to sprouting in garlic is not more than *(ARS, 2003)*
(a) 0.1% (b) 0.2%
(c) 0.3% (d) 0.5%

61. Onion belongs to the family *(ARS, 1992)*
(a) *Liliaceae* (b) *Araceae*
(c) *Amaryllidaceae* (d) *Solanaceae*

62. Onion is propagated through *(ARS, 1992)*
(a) Seed (b) Cloves
(c) Roots (d) Stem

63. Male sterility in onion was first located by H.A. Jones in the year *(ARS, 1992)*
(a) 1925 (b) 1935
(c) 1995 (d) 1965

64. Garlic belongs to the species *(Ph.D. VGS, 2006)*
(a) *cepa* (b) *sativum*
(c) *porrum* (d) *carota*

65. The onion variety which is grown during *Kharif* is *(UPPCS (Pre), 1998)*
(a) N-53 (b) Kalyanpur Red Round
(c) Pusa Madhavi (d) Early Grano

66. The onion variety suitable for *kharif* season in U.P. is *(UP PCS (Pre), 1998, 2000, 2002)*
(a) Agrifound Red (b) Agrifound Light Red
(c) Agrifound Dark Red (d) Punjab Red Round

67. Agrifound Red is a variety of *(ARS, 1999)*
(a) Garlic (b) Tomato
(c) Onion (d) Pea

68. Optimum temperature for onion crop is *(ARS, 1999)*
(a) 30°C (b) 25°C
(c) 20°C (d) 15°C

69. The yellow colour in onion is due to the pigment *(Ph.D. VGS, 2004)*
   (a) Anthocyanin (b) Quercetin
   (c) Lycopene (d) Carotene
70. Purple blotch disease of onion is caused by *(Ph.D. VGS, 2005)*
   (a) Aphids (b) Thrips
   (c) Whitefly (d) Nematodes
71. Garlic belongs to be natural order *(UP PCS (Pre), 2006)*
   (a) Apiaceae (b) Liliaceae
   (c) Amaryllidaceae (d) Alliaceae
72. Herbicide suitable for weed control in onion is *(UP PCS (Pre), 2006)*
   (a) Karmax (b) Sencore
   (c) Goal (d) Basalin
73. Which of the following is used to check the sprouting of onion during storage ?
   (a) NAA (b) GA
   (c) MH (d) PCPA
74. Pre-harvest foliage spray of Maleic hydrazide is helpful in controlling the sprouting in storage of *(UP PCS, 2005)*
   (a) Potato (b) Onion
   (c) Turnip (d) Sweet potato

ANSWERS

1. (b)	2. (a)	3. (c)	4. (c)	5. (a)
6. (b)	7. (b)	8. (b)	9. (c)	10. (c)
11. (b)	12. (b)	13. (a)	14. (a)	15. (d)
16. (c)	17. (c)	18. (c)	19. (a)	20. (b)
21. (a)	22. (c)	23. (c)	24. (b)	25. (b)
26. (b)	27. (c)	28. (c)	29. (b)	30. (d)
31. (a)	32. (b)	33. (b)	34. (a)	35. (b)
36. (b)	37. (b)	38. (c)	39. (c)	40. (a)
41. (a)	42. (a)	43. (b)	44. (a)	45. (b)
46. (b)	47. (a)	48. (b)	49. (c)	50. (a)
51. (c)	52. (c)	53. (a)	54. (d)	55. (c)
56. (c)	57. (b)	58. (c)	59. (c)	60. (d)
61. (c)	62. (a)	63. (a)	64. (b)	65. (a)
66. (c)	67. (c)	68. (b)	69. (b)	70. (b)
71. (d)	72. (d)	73. (c)	74. (b)	

CHAPTER 8

ROOT CROPS

Radish, turnip, sugar beet and carrot are important root vegetables.

8.1 Radish

- Radish belonging to family *Cruciferae* is a root-cum-leafy vegetable suitable for tropical and temperate climate.
- The edible portion of radish root develops from both primary root and the hypocotyl.
- Radish is a cross-pollinated vegetable due to sporophytic system of self-incompatibility.
- Pusa Chetki variety is suitable for growing in hotter months, i.e. middle of March to middle of August when no other varieties can be grown successfully.
- Arka Nishant is resistant to pithiness, pre-mature bolting, root branching and forking.
- Pusa Himani is a cross between Black and Japanese White.
- Pusa Himani is the only variety which can be grown throughout the year.
- Wart is a physiological disorder which is a white protrusion of white inner root tissue mainly due to soil moisture deficit.
- Hollow root is a physiological disorder which is due to high temperature during 16-30 days of sowing.

8.2 Carrot

- Carrot is another root crop of *umbelliferae* family which is an annual herb for root production and biennial for flowering and seed set.
- Carrot flower is protandrous. The inflorescence is a compound umbel.
- The temperate types form roots under both tropical and temperate conditions but set seeds only under temperate conditions as they need low temperature of 5-8°C for 40-60 days before throwing out the seed stalks.
- The Asiatic types are high yielding and produce seeds under tropical conditions. But they are poor in carotene content.

- Pusa Kesar is a selection from a cross of Local Red and Nantes Half Long.
- Pusa Meghali is a selection from a cross between Pusa Kesar and Nantes.
- Chantaney is an excellent cultivar for canning and storage.
- Cavity spot is a physiological disorder of carrot due to calcium deficiency.

8.3 Beetroot

Beetroot is a root crop of *chenopodiaceae* family. The probable origin of beetroot is Mediterranean region for *Beta vulgaris* L. ssp. *Maritimat,* possibly by hybridization with *B. patula.*

- The colour of root is due to the presence of red-violet pigment, *B*-cyanin and yellow pigment, *B*-xanthins.
- The inflorescence is large spike, which normally develops in the second year.
- A temperature of 4.5°C-10°C for 15 days or more may induce premature bolting in beetroot.
- Viral diseases like beet mosaic yellow are transmitted by aphids and curly top by hoppers.
- Internal black spot or brown heart rot of beetroot is caused due to boron deficiency.
- The beets are harvested when they attain a diameter of 3-5 cm.

8.4 Turnip

Turnip is an important root vegetable of *Cruciferae* family grown as a summer crop in temperate climate and as a winter vegetable in subtropical places where the winter is not severe.

- The fleshy thickened underground portion of turnip is actually the hypocotyl.
- A distinct tap root and secondary roots of turnip arise from the lower part of the swollen hypocotyl.
- Normally, the roots of turnip attain edible maturity in 40-80 days.
- The inflorescence is a terminal raceme on the main stem.
- Turnip has sporophytic system of self-incompatibility.
- Pusa Swarnima was developed by hybridization and selection by involving Asiatic (Japanese white) and European (Golden Ball) types.
- Pusa Kanchan is a selection from the cross between Local Red Round (Asiatic) and Golden Ball (European).
- Pusa Kanchan also sets seeds in the plains.

Objective type questions on Root crops are given below :

1. *Raphanus sativus* L. is the botanical name of
   (a) Turnip (b) Radish
   (c) Cabbage (d) None of these
2. Family of radish is
   (a) *Chenopodiaceae* (b) *Cruciferae*
   (c) Both (d) None of these
3. Edible portion of radish is
   (a) Primary root (b) Hypocotyle
   (c) Both (d) None of these
4. Radish is a
   (a) Self-pollinated crop (b) Cross-pollinated crop
   (c) Both (d) None of these
5. Radish variety suitable for summer season is
   (a) Pusa Chetki (b) Pusa Himani
   (c) Both (d) None of these
6. Arka Nishant variety of radish is resistant to
   (a) Pithiness (b) Pre-mature bolting
   (c) Forking (d) All the above
7. Pusa Himani is a variety developed through a cross between 'Black' and Japanese
   (a) Yellow (b) White
   (c) Purple (d) None of these
8. Radish variety which may be grown throughout the year is
   (a) Pusa Himani (b) Arka Nishant
   (c) Pusa Chetki (d) Kalyanpur No. 1
9. Akashin of radish is a physiological disorder due to deficiency of
   (a) Calcium (b) Boron
   (c) Nitrogen (d) None of these
10. Excess application of NPK and soil moisture stress in radish results in
    (a) Pithiness (b) Forking
    (c) Both (d) None of these
11. Wart is a physiological disorder of
    (a) Radish (b) Carrot
    (c) Both (d) None of these
12. Hollow-root is a physiological disorder due to
    (a) High temperature (b) Low temperature
    (c) High humidity (d) None of these

13. *Daucus carota* L. is the botanical name of
   (a) Carrot (b) Radish
   (c) Turnip (d) None of these
14. Chromosome number of *Daucus carota* is
   (a) 2n = 14 (b) 2n = 16
   (c) 2n =18 (d) None of these
15. Seed rate (kg/ha) of carrot is
   (a) 2-4 (b) 4-6
   (c) 6-8 (d) None of these
16. Seed rate (kg/ha) of radish is
   (a) 2-4 (b) 4-6
   (c) 6-8 (d) 8-10
17. The inflorescence of carrot is
   (a) Umbel (b) Compound Umbel
   (c) Both (d) None of these
18. Seed setting temperature for carrot is
   (a) 5-8°C (b) 5-10°C
   (c) 10-12°C (d) None of these
19. The Asiatic type varieties of carrot are high yielding and produce seeds under
   (a) Temperate conditions (b) Tropical conditions
   (c) Both (d) None of these
20. Pusa Kesar is a selection from a cross Local Red and--------
   (a) Nantes (b) Nantes Half long
   (c) Asiatic Long (d) None of these
21. Pusa Meghali is a selection from cross Pusa Kesar and-------
   (a) Nantes (b) Nantes Half Long
   (c) Asiatic long (d) None of these
22. The cultivar of carrot suitable for canning and storage purpose is
   (a) Pusa Kesar (b) Pusa Meghali
   (c) Chantancy (d) None of these
23. Splitting of carrot root is due to
   (a) Low dose of Nitrogen (b) Heavy dose of Nitrogen
   (c) Calcium deficiency (d) None of these
24. *Beta vulgaris* L. var *rubra* is the botanical name of
   (a) Radish (b) Carrot
   (c) Beetroot (d) None of these
25. Beetroot is also known as
   (a) Sugar beet (b) Chukandar
   (c) Both (d) None of these

26. Family of sugar beet is
    (a) *Chenopodiaceae* (b) *Cucurbitaceae*
    (c) *Leguminosae* (d) None of these
27. Origin of sugar beet is
    (a) Asia (b) India
    (c) Mediterranean region of Europe
    (d) None of these
28. Red colour of beet root is due to
    (a) β-Cyanins (b) β- xanthins
    (c) Both (d) None of these
29. Yellow colour of beetroot is due to
    (a) β-Cyanins (b) β-xanthins
    (c) Both (d) None of these
30. Beet mosaic disease is transmitted by
    (a) Aphids (b) White fly
    (c) Both (d) None of these
31. Heart rot of beetroot is caused due to deficiency of
    (a) Calcium (b) Boron
    (c) Both (d) None of these
32. The sugar beet is harvested when the roots attain a diameter of
    (a) 2-3 cm (b) 1-2 cm
    (c) 4-5 cm (d) None of these
33. *Brassica campestris* sp. *rapa* is the botanical name of
    (a) Radish (b) Cabbage
    (c) Turnip (d) None of these
34. Chromosome number of turnip is
    (a) 2n = 18 (b) 2n = 20
    (c) 2n = 24 (d) None of these
35. Seed rate (kg/ha) of turnip is
    (a) 1-1·5 (b) 2.5 – 2.8
    (c) 2.5 –3.5 (d) None of these
36. Edible part of turnip is
    (a) Hypocotyle (b) Swollen hypocotyle
    (c) Stem (d) None of these
37. The edible maturity of turnip is
    (a) 30 days (b) 40 days
    (c) 40-80 days (d) None of these
38. Inflorescence of turnip is
    (a) Racemose (b) Cymose
    (c) Both (d) None of these

39. Self-incompatibility in turnip is
(a) Gametophytic (b) Sporophytic
(c) Both (d) None of these

40. Pusa Swarnima, Japanese White and Golden Ball are the improved varieties of
(a) Radish (b) Carrot
(c) Beet root (d) Turnip

41. Pusa Kanchan is an improved variety of
(a) Radish (b) Turnip
(c) Both (d) None of these

42. Seed production of Pusa Kanchan can be done in
(a) Hills (b) Plains
(c) Both (d) None of these

43. Extra early and high yielding cultivar of turnip is
(a) Pusa Kanchan (b) Pusa Swarnima
(c) Early Milan Red Top (d) None of these

44. Turnip yellow mosaic virus disease is transmitted through
(a) Aphids (b) White fly
(c) Flea beetle (d) None of these

45. Delayed harvesting in radish may cause *(ARS, 2003)*
(a) Forking (b) Pithiness
(c) Deformed roots (d) All of these

46. The most common insect attacking radish is *(ARS, 2003)*
(a) Red pumpkin bettle (b) Aphids
(c) Thrips (d) None of these

47. The most commonly grown garden beet variety is *(ARS, 2003)*
(a) Detriot Dark Red (b) Pusa Kanchan
(c) Pusa Chandrima (d) None of these

48. Carrot belongs to the family *(ARS, 1992)*
(a) *Solanaceae* (b) *Cruciferae*
(c) *Umbelliferae* (d) *Malvaceae*

49. Pusa Kesar is the variety of *(ARS, 1992)*
(a) Radish (b) Turnip
(c) Beetroot (d) Carrot

50. The tropical type variety of radish is *(ARS, 1992)*
(a) Pusa Chetki (b) White Icicle
(c) Scarlet Globe (d) None of these

51. The botanical name of radish is *(ARS, 1992)*
(a) *Daucus carota* (b) *Raphanus sativus*
(c) *Pisum sativum* (d) *Brassica rapa*

52. The pungency in radish is due to *(ARS, 1992)*
    (a) Capsanthin (b) Sulphur
    (c) Isothiocyanates (d) None of these

53. Pink colour in skin of the radish root is due to *(ARS, 1992)*
    (a) Anthocyanin (b) Capsaicin
    (c) Carotenoid (d) Chlorophyll

54. Radish is a *(ARS, 1992)*
    (a) Cross-pollinated (b) Self-pollinated
    (c) Often cross-pollinated (d) None of the above

55. Brown heart is a physiological disorder in turnip which is caused by deficiency of *(ARS, 1992)*
    (a) Ca (b) Mo
    (c) B (d) Zn

56. For colour development the optimum temperature for carrot crop is *(ARS, 1999)*
    (a) 5-10°C (b) 10-13°C
    (c) 21-35°C (d) None of these

57. The origin of beetroot is *(ARS, 1999)*
    (a) Russia (b) South Africa
    (c) North America (d) Mediterranean

58. Pusa Yamdagini is a variety of *(ARS, 1999)*
    (a) Radish (b) Turnip
    (c) Carrot (d) None of these

59. Pusa Sweti variety of turnip was developed through *(ARS, 1999)*
    (a) Mutation (b) Inbreeding
    (c) Mass Selection
    (d) Hybridization followed by pedigree selection

60. Pusa Kesar carrot was developed from the cross *(ARS, 1999)*
    (a) Nantes x Pusa Desi (b) Nantes x Chantaney
    (c) Nantes x Local Asiatic type
    (d) China Red x Sel. 5

61. Sugar beet matures in *(Ph.D. VGS, 2003)*
    (a) April/May (b) June/July
    (c) September/October (d) January/February

62. For ideal plant population, the seed rate of sugar beet should be (kg/ha) *(Ph.D. VGS, 2003)*
    (a) 4-8 (b) 8-10
    (c) 10-15 (d) 25-30

63. Sugar beet belongs to the family
    (a) *Graminae* (b) *Convolvulaceae*
    (c) *Chenopodiaceae* (d) None of these

64. Sugar beet is an indicator plant for deficiency of *(Ph.D. VGS, 2004)*
    (a) Phosphorus (b) Magnesium
    (c) Boron (d) Sodium

65. Carrot is a heavy feeder particularly for
    (a) N (b) P
    (c) K (d) Ca

66. Pusa Chetki Variety of radish is most suitable for sowing during *(ARS, 2003)*
    (a) March-August (b) October-November
    (c) After November (d) None of the above

67. Tropical carrots are sown in the plains from *(ARS, 2003)*
    (a) Mid August to beginning of September
    (b) November-December
    (c) July- August
    (d) None of the above

68. Carrot is the main source of *(UP PCS, 2005)*
    (a) Carotene (b) Protein
    (c) Pectin (d) Vitamin C

ANSWERS

1. (b)	2. (b)	3. (c)	4. (b)	5. (a)
6. (d)	7. (b)	8. (a)	9. (b)	10. (a)
11. (a)	12. (a)	13. (a)	14. (c)	15. (c)
16. (d)	17. (b)	18. (a)	19. (b)	20. (b)
21. (a)	22. (c)	23. (b)	24. (c)	25. (c)
26. (a)	27. (c)	28. (a)	29. (b)	30. (a)
31. (b)	32. (c)	33. (c)	34. (b)	35. (c)
36. (b)	37. (c)	38. (a)	39. (b)	40. (d)
41. (b)	42. (b)	43. (c)	44. (c)	45. (b)
46. (b)	47. (a)	48. (c)	49. (d)	50. (a)
51. (b)	52. (c)	53. (a)	54. (a)	55. (c)
56. (b)	57. (d)	58. (c)	59. (d)	60. (d)
61. (a)	62. (b)	63. (c)	64. (d)	65. (c)
66. (b)	67. (a)	68. (a)		

CHAPTER 9

SOLANACEOUS VEGETABLES

Solanaceae is a very large vegetable family. The major vegetables like potato, tomato, brinjal and chillies belong to this family. Potato is a major vegetable having high bulking and yielding potential.

9.1 Tomato

- Tomato, universally treated as protective food, is being extensively grown as annual plant all over the world.
- The specific requirements of tomato for export in West Asia are round, medium size and red fruit, while cherry tomato is preferred for export to European countries.
- The export quality tomato producing areas in India are Nasik and Pune in Maharashtra and Bangalore in Karnataka.
- Processed products of tomato, especially 'Tomato Puree'.and 'Tomato Paste' have great demand for export.
- Growth habit of tomato plant is (1) *determinate* – plants are dwarf and growth is restricted with the appearance of terminal flower; (2) *indeterminate* – continuous growth and there is less initiation of flowers and fruits on the stem.
- Lycopene, which is responsible for red colour in tomato, is highest at 21°C – 24°C, while the production of this pigment drops rapidly when temperature exceeds 27°C.
- For early crop of tomato, a sandy loam soil is the best. However, for highest yield, heavy soils rich in organic matter are preferred.
- Tomato is moderately tolerant to acid soil (pH 5.5).
- Varieties Pusa Early Dwarf, Pusa Ruby, Arka Vikas, Arka Saurabh, Punjab Kesari and Pant Bahar are suitable for fresh market.
- Varieties Pusa Gaurav, Roma, Punjab Chhuhara, Pusa Uphar, Arka Saurabh are suitable for processing.
- Pusa Sheetal is suitable for low temperature regions.
- Pusa Hybrid 1 is suitable for high temperature regions.
- Pusa Sadabahar is suitable for both high and low temperature regions.

- Sel-120 is the first Root-knot nematode resistant variety of tomato.
- Seed treatment of tomato with 2, 4-D @ 2-5 ppm increases fruit set and induces earliness and parthenocarpy.
- Pruning and training are generally followed in indeterminate tomatoes.
- Tomato is climacteric fruit.
- In order to enhance the ripening of tomato fruits, ethrel (1,000 ppm) can be sprayed on the plants at the time of initiation of ripening.
- In 1992, Greshof and Doy produced haploid plants of tomato by anther culture.

9.2 Brinjal

Brinjal also known as egg plant or aubergine is another important vegetable of *Solanaceae* family family and is a day-neutral plant.

- Brinjal fruits are good source of vitamin B.
- Pigmented and dark-purple brinjal has more vitamin C than those with white skin.
- Brinjal flowers are pentamerous, hermaphrodite and solitary with self-pollination but cross-pollination also occurs. This is because it has pronounced heterostyly which favours cross-pollination.
- In brinjal, opening of flowers takes place at 6-7.30 a.m. in summer which is delayed to 11:15 a.m. in winter.
- Brinjal flowers are divided into four types, namely (i) long styled, (ii) medium styled, (iii) pseudo-short styled and (iv) short styled.
- The maximum fruit setting in different cultivars is in 'long styled flowers' which ranges from 70-80%. In medium styled, it ranges between 12 and 55%. There is no fruit setting in pseudo-short and short styled flowers.
- Brinjal cultivars belong to three categories of varieties (a) *esculentum,* which includes round and egg-shaped cultivars; (b) *serpentinum;* long, slender types; and (c) *depressum,* which includes all early and dwarf cultivars.
- Pusa Purple Long is an extra early cultivar.
- Arka Shirish is green coloured cultivar.
- KKM-1 is milky white coloured cultivar.
- Pusa Bhairav and Pusa Anupam are resistant to *Phomopsis* fruit rot.
- Pusa Anupam is a cross between PPC and Pusa Kranti.
- Pragati is a cross between Vaishali and Manjari Gota.
- Brinjal green fruits have the longest shelf life of 4 weeks.

9.3 Chilli

- Chilli is known as 'hot pepper' and capsicum as 'bell pepper'.
- Andhra Pradesh has been leading both in area and production of chilli.
- In food and beverage industries, chilli is being used in the form of oleoresin which permits better distribution of colour and flavour in food.
- *Capsicum* is the pungent principle found in chilli.
- *C. annuum* and *C. frutesceuns* have 'white flowers' while *C. pubesceuns* has 'purple flowers'.
- Chilli and capsicum can be grouped in often-cross-pollinated group. Cross pollination is reported upto 63%.
- Majority of the flowers open at 5 a.m.
- MDU 1 is a mutant (irradiation) variety.
- Sindhur is mild pungent variety.
- Jwalamukhi is a variety suitable for high-density planting, has low pungency and is resistant to virus.
- Punjab Lal is resistant to TMV, CMV, leaf curl virus, fruit rot and die back.
- CH1 is F_1 hybrid (using GMS).
- Pusa Deepti is F_1 hybrid of capsicum.

The objective type questions on *solanaceous* vegetables are given below :

1. 'Protective' food crop is
   (a) Potato (b) Tomato
   (c) Brinjal (d) None of these
2. Cherry tomato is preferred for export to
   (a) Asia (b) West Asia
   (c) European countries (d) None of these
3. Medium size, round and red coloured fruits of tomato are preferred for export to
   (a) Africa (b) West Asia
   (c) European countries (d) None of these
4. The export quality tomato producing areas are
   (a) Maharashtra (b) Karnataka
   (c) Both (d) None of these
5. Processed products for export of tomato is
   (a) Tomato paste (b) Tomato puree
   (c) Both (d) None of these
6. Native of tomato is
   (a) West America (b) America
   (c) South America (d) None of these

7. Tomato has
   (a) Tap root system (b) Shallow root system
   (c) Both (d) None of these
8. Growth habit of tomato is
   (a) Determinate (b) Indeterminate
   (c) Both (d) None of these
9. Red colour in tomato is highest at the temperature of
   (a) 15-18°C (b) 21-24°C
   (c) >24°C (d) None of these
10. Tomato produces Lycopene pigment which drops rapidly at the temperature of
    (a) >15°C (b) >20°C
    (c) >24°C (d) >27°C
11. For growing early crop of tomato, the best soil is
    (a) Sandy loam (b) Clay loam
    (c) Both (d) None of these
12. For higher yield of tomato, the soil should be
    (a) Heavy (b) Clay
    (c) Clay loam (d) Sandy
13. Tomato is moderately tolerant to
    (a) Acidic Soil (pH 5.5) (b) Alkaline soils
    (c) Non-acidic soils (d) None of these
14. Tomato varieties suitable for fresh market are
    (a) Pusa Ruby (b) Pusa Early Dwarf
    (c) Punjab Kesari and Pant Bahar Azad T-5
    (d) All the above
15. Tomato varieties suitable for processing are
    (a) Pusa Gaurav, Pusa Uphar (b) Arka Saurabh
    (c) Roma (d) All the above
16. Tomato variety suitable for low temperature regions is
    (a) Pusa Gaurav (b) Pusa Sheetal
    (c) Both (d) None of these
17. Tomato variety suitable for high temperature is
    (a) Pusa Hybrid-1 (b) Pusa Uphar
    (c) Both (d) None of these
18. Tomato variety suitable for both low and high temperature regions is
    (a) Pusa Sadabahar (b) Pusa Gaurav
    (c) Arka Saurabh (d) None of these

19. Seed rate of hybrid tomato per ha is
    (a) 50-60 g (b) 60-100 g
    (c) 125-175 g (d) None of these
20. Seedlings of tomato soaked for 24 hours in NAA at 0.1 ppm show
    (a) Early flowering (b) Highest fruit setting
    (c) Both (d) None of these
21. Seed treatment of tomato with 2,4-D @ 2-5 ppm increases
    (a) Fruit set (b) Earliness
    (c) Parthenocarpy (d) All the above
22. Tomato is a
    (a) Climacteric fruit (b) Non-climacteric fruit
    (c) Both (d) None of these
23. The ripening of tomato fruits Ethrel can be sprayed at the time of initiation of ripening with concentration of
    (a) 100 ppm (b) 500 ppm
    (c) 1000 ppm (d) None of these
24. The first Root-Knot nematode resistant variety of tomato is
    (a) Sel-120 (b) Hissar Lalit
    (c) Both (d) None of these
25. Nematode and Bacterial wilt resistant variety of tomato is
    (a) Sel-120 (b) Hissar lalit
    (c) Arka Vardan (d) All the above
26. In 1992, Greshof and Doy produced haploid plants of tomato by
    (a) Tissue culture (b) Anther culture
    (c) Both (d) None of these
27. Brinjal is also known as
    (a) Egg plant (b) Aubergine
    (c) Both (d) None of these
28. Brinjal fruits are good source of
    (a) Vit-A (b) Vit-B
    (c) Vit-C (d) None of these
29. Dark-purple brinjal has more vitamin .... that those with white skin
    (a) A (b) B
    (c) C (d) None of these
30. Brinjal flowers are
    (a) Pentamerous (b) Hermaphrodite
    (c) Solitary (d) All the above

31. In brinjal heterostyly is common feature except in
    (a) Long type cultivars (b) Solitary type cultivars
    (c) Bunchy type cultivars (d) None of these
32. Brinjal fruit is
    (a) Pome (b) Berry
    (c) Both (d) None of these
33. Brinjal is a
    (a) Long-day plant (b) Short-day plant
    (c) Day-neutral plant (d) None of these
34. Optimum temperature for normal growth and development of brinjal plant and fruits is
    (a) 21-30°C (b) 30-32°C
    (c) 32-40°C (d) None of these
35. The fruit setting range in long styled flowers of brinjal is
    (a) 40-50% (b) 50-60%
    (c) 60-70% (d) 70-80%
36. The fruit setting range in medium styled flowers of brinjal is
    (a) 12-55% (b) 55-60%
    (c) 60-70% (d) None of these
37. The fruit setting range in Pseudo-short styled and short styled brinjal is
    (a) 12-55% (b) 55-60%
    (c) 60-70% (d) No setting
38. Round and egg-shaped brinjal cultivars belong to the variety group
    (a) Esculentum (b) Serpentinum
    (c) Depressum (d) None of these
39. Long slender type brinjal cultivars belong to variety group
    (a) Esculentum (b) Serpentinum
    (c) Depressum (d) None of these
40. Early and dwarf type brinjal cultivars belong to variety group
    (a) Esculentum (b) Serpentinum
    (c) Depressum (d) None of these
41. Extra early maturity cultivar of brinjal is
    (a) Pusa Purple Long (b) Pant Samrat
    (c) Both (d) None of these
42. Aphid resistant variety of brinjal is
    (a) Pant Samrat (b) Annamalai
    (c) Both (d) None of these

43. Green coloured cultivar of brinjal is
    (a) Arka Shirish (b) Arka Vikas
    (c) Both (d) None of these
44. Milky white coloured cultivar of brinjal is
    (a) Arka Shirish (b) Annamalai
    (c) KKM-1 (d) None of these
45. Phomopsis fruit rot resistant variety of brinjal is
    (a) Pusa Bhairav (b) Pusa Anupam
    (c) Both (d) None of these
46. Pusa Anupam cultivar of brinjal is a cross between
    (a) PPC and Pusa Kranti (b) PPC and PPL
    (c) PPC and KKM-1 (d) None of these
47. Vaishali cultivar of brinjal is a cross between
    (a) Arka Kusumakar and Manjari Gota
    (b) Arka Kusumakar and Pusa Kranti
    (c) PPC and PPL (d) None of these
48. Pragati cultivar of brinjal is a cross between
    (a) Arka Kusumakar and Manjari Gota
    (b) Vaishali and Manjari Gota
    (c) PPC and PPL (d) None of these
49. The soil suitable for early production of brinjal is
    (a) Sandy (b) Silt loam
    (c) Clay loam (d) All the above
50. The soil suitable for maximum production of brinjal is
    (a) Silt loam (b) Clay loam
    (c) Both (d) None of these
51. The number of seedlings of brinjal sufficient to cover one hectare of field is
    (a) 30,000-45,000 (b) 45,000-50,000
    (c) 50,000-60,000 (d) None of these
52. In brinjal application of copper and manganese increases
    (a) Number of flowers (b) Number of fruits
    (c) Both (d) None of these
53. In brinjal application of zinc improves
    (a) Number of flowers (b) Number of fruits
    (c) Fruit weight (d) None of these
54. Root parasitic weed of brinjal is
    (a) Pigweed (b) Orobanchae
    (c) Both (d) None of these
55. The irrigation required for normal brinjal crop is
    (a) 50-60 cm (b) 60-100 cm
    (c) 100-110 cm (d) None of these

56. In brinjal application of 2,4-D (2 ppm) at flowering induces
(a) Parthenocarpy (b) Fruit setting
(c) Timely fruit maturation (d) All the above

57. In brinjal, the highest yield is obtained from plants whose roots are dipped in 250 ppm solution of
(a) GA_3 + ascorbic acid (b) GA_3 +NA
(c) Both (d) None of these

58. The longest shelf life of brinjal fruit has the colour
(a) White (b) Green
(c) Both (d) None of these

59. Little leaf resistant variety of brinjal is
(a) Arka Sheel (b) Manjari Gota
(c) Both (d) None of these

60. Chilli is also known as
(a) Bell pepper (b) Hot pepper
(c) Both (d) None of these

61. The state having highest area and production of chilli is
(a) Andhra Pradesh (b) Uttar Pradesh
(c) Madhya Pradesh (d) None of these

62. Pungency in chilli is due to
(a) Solanin (b) Capsicin
(c) Both (d) None of these

63. The Portuguese brought capsicum from Brazil to India during
(a) 1564 (b) 1574
(c) 1584 (d) None of these

64. *Capsicum annuum* and *C. frutesceuns* have white flowers, while *C. pubesceuns* has flowers of
(a) Yellow (b) Purple
(c) White (d) None of these

65. Often-cross pollinated crop is
(a) Chilli (b) Capsicum
(c) Both (d) None of these

66. In capsicum and chilli the cross pollination is reported upto
(a) 30% (b) 48%
(c) 50% (d) 63%

67. In chilli majority of flowers open at
(a) 4 a.m. (b) 5 a.m.
(c) 6 a.m. (d) None of these

68. MDU-1 is a mutant variety of
(a) Okra (b) Chilli
(c) Capsicum (d) None of these

69. Mild pungent variety of chilli is
    (a) Jwalamukhi (b) Sindhur
    (c) Both (d) None of these
70. Variety of chilli suitable for high-density planting and low pungency and resistant to virus is
    (a) Jwalamukhi (b) Sindhur
    (c) Punjab Lal (d) None of these
71. Variety of chilli resistant to TMV, CMV, LCV, fruit rot and die back is
    (a) Jwalamukhi (b) Sindhur
    (c) Punjab Lal (d) None of these
72. Chilli variety suitable for colour extraction is
    (a) Jwalamukhi (b) Sindhur
    (c) Punjab Lal (d) Arka Abir
73. F_1 hybrid of chilli is
    (a) CH-1 (b) CH-2
    (c) Pusa Deepti (d) None of these
74. F_1 Hybrid of capsicum is
    (a) CH-1 (b) Pusa Deepti
    (c) Both (d) None of these
75. The number of seedlings of chilli sufficient to cover one hectare of field is
    (a) 40,000-50,000 (b) 30,000-45,000
    (c) 50,000-60,000 (d) None of these
76. Fruit setting in chilli can be improved by the application of
    (a) GA (10-100 ppm) (b) NAA (20-200 ppm)
    (c) CCC (1000 ppm) (d) All the above
77. Fruit setting in brinjal is usually high in the flower having *(ARS 2003)*
    (a) Long and medium style (b) Short and medium style
    (c) Medium style (d) Short style
78. Pruning and training of tomato plants to a single stem help in getting *(ARS, 2003)*
    (a) High total yield (b) High marketable yield
    (c) High early yield (d) Increased fruit size
79. Puffy fruit in tomato is caused by *(ARS, 2003)*
    (a) Viral attack (b) Physiological disorder
    (c) Bacterial attack (d) None of these
80. Very important vegetable included in solanaceous fruit group is *(ARS, 2003)*
    (a) Tomato (b) Potato
    (c) Okra (d) Sweet potato

81. Virus diseases of chilli are spread by *(ARS, 2003)*
(a) White flies (b) Thrips
(c) Aphids (d) None of these

82. Chilli hybrid-1 CH-1 was evolved by *(ARS, 2003)*
(a) CSAU and T (Kanpur) (b) PAU (Ludhiana)
(c) CCHAU (Hissar) (d) IARI (New Delhi)

83. Chillies can be grown from sea level to an altitude of *(ARS, 2003)*
(a) 1000 m (b) 2000 m
(c) 3000 m (d) 4000 m

84. Chilli Guchchadar is suitable for *(ARS, 2003)*
(a) Drying and canning (b) Salad purpose
(c) Drying only (d) None of these

85. Tomato variety RKMT has been evolved through *(ARS, 2003)*
(a) X-rays (b) Gamma rays
(c) EMS (d) MMS

86. Pollination and fertilization processes in tomato are impaired due to night temperature below *(ARS, 2003)*
(a) 13°C (b) 5°C
(c) 10°C (d) 20°C

87. Heterosis in brinjal was studied by *(ARS, 2003)*
(a) B.S. Tomar (b) H. Singh and T.S. Kaider
(c) N. Basuvaraja (d) All of the above

88. For canning tomato fruits are picked at *(ARS, 2003)*
(a) Immature green stage (b) Mature green stage
(c) Half ripe or pink stage (d) Red ripe stage

89. SL-120 semi determinate variety of tomato was evolved by *(ARS, 2003)*
(a) GB PUA T, Pantnagar (b) PAU, Ludhiana
(c) IARI, New Delhi (d) IIHR, Bangalore

90. During cropping period of brinjal the average number of irrigations required is *(ARS, 2003)*
(a) 5 (b) 8
(c) 16 (d) None of these

91. Haploid plants have been produced by anther culture in tomato by *(ARS, 2003)*
(a) Greshop and Doy (1972)
(b) Tyagi and Prasad (1965)
(c) Mehta and Upadhyaya (1960)
(d) H.B. Singh (1963)

92. The best temperature for storing tomato fruits is *(ARS, 1992)*
    (a) 5-10 °C (b) 10-12 °C
    (c) 12-15 °C (d) 15-20 °C
93. The variety of brinjal is *(UP PCS (Pre), 1994)*
    (a) Pusa Kesar (b) Pusa Ruby
    (c) Pusa Kranti (d) Pusa Chetki
94. Origin of brinjal is in *(UP PCS (Pre), 1995)*
    (a) India (b) Africa
    (c) America (d) China
95. Fruit and shoot borer is a serious pest of *(Ph.D. VGS, 2006)*
    (a) Brinjal (b) Tomato
    (c) Bhindi (d) Cabbage
96. In chilli fruit drop is controlled by the spray of *(ARS. (Pre), 2003)*
    (a) NAA 10 ppm (b) NAA 20 ppm
    (c) NAA 40 pppm (d) NAA 50 ppm
97. Seed rate of tomato is g/ha *(UP PCS (Per), 1996)*
    (a) 400-500 (b) 1200-1500
    (c) 800-1000 (d) 100-200
98. Pollinating agent in chilli is *(JRF, 1998)*
    (a) House fly (b) Bee
    (c) Wind (d) Water
99. For chilli optimum pH range is *(ARS (Pre), 2003)*
    (a) 4.5-5.5 (b) 5.8-6.5
    (c) 6.0-7.0 (d) 6.5 to 7.5
100. Tomato is considered to be *(Ph.D. VGS, 2006)*
    (a) Day neutral plant (b) Long-day plant
    (c) Short-day plant (d) None of these
101. The variety of brinjal having round fruits is *(Ph.D. VGS, 2006)*
    (a) Pant Samrat (b) Pant Rituraj
    (c) Pusa purple cluster (d) None of the above
102. Pant C is natural cross between the parents *(ARS (Pre), 2003)*
    (a) Pusa Jwala x NP 46-A (b) NP 46-A x Chanobal
    (c) NP 46-A x Local Kandhari
    (d) Local Kandhari x Pusa Jwala
103. For canning the proper stage of harvesting tomato is *(Ph.D. VGS, 2006)*
    (a) Pink stage (b) Mature green
    (c) Turning red (d) Red ripe

104. Which is not the variety of chilli *(ARS (Pre), 2003)*
(a) Pusa Sadabahar (b) NP 46-A
(c) Andhra Jyoti (d) Arka Mohini

105. The following vegetable group classified as solanaceous fruits is *(Ph.D. VGS, 2006)*
(a) Peas, beans (b) Lettuce, celery, parsley
(c) Tomato, brinjal, chillies (d) Cucumber, squash, okra

106. The red colour of tomato is due to *(UP PCS (Pre), 1998)*
(a) Chromoplast (b) β-carotene
(c) Lycopene (d) Anthocyanin

107. S-12 Tomato variety is the *(ARS, 1999)*
(a) Hybridization (b) Mass selection
(c) Pure line selection (d) Introduction

108. The bitter taste in brinjal fruits is due to *(ARS, 1999)*
(a) Marmelocin (b) Sulphur
(c) Borane (d) Glycoalkaloids

109. Little leaf disease of brinjal is caused by *(ARS, 1999)*
(a) Phytoplasma (b) Mycoplasma
(c) Virus (d) None of these

110. The best pH for tomato cultivation is *(ARS, 1999)*
(a) 4-5 (b) 5-6
(c) 6-7 (d) 7-8

111. The chromosome number of tomato is *(ARS, 1999)*
(a) 2n = 48 (b) 2n = 24
(c) 2n = 42 (d) None of these

112. Brinjal belongs to the species *(ARS, 1999)*
(a) *esculentus* (b) *melongena*
(c) *khasianum* (d) None of these

113. Blossom end rot in tomato is caused by the deficiency of *(ARS, 1999)*
(a) Ca (b) B
(c) Mg (d) Cl

114. Tomato is considered to be *(ARS, 1999)*
(a) Long-day (b) Day-neutral
(c) Short-day (d) None of these

115. Mulching in brinjal is recommended with *(ARS, 1999)*
(a) Straw (b) Sand
(c) White polythene film (d) Black polythene film

116. The isolation distance for producting certified seed of brinjal is *(ARS, 1999)*
(a) 100 m (b) 200 m
(c) 300 m (d) 400 m

117. The number of types of flowers found in brinjal is *(ARS, 1999)*
   - (a) Four (b) Five
   - (c) Six (d) Seven
118. The preservative used for preparation of tomato ketchup is *(Ph.D. VGS, 2004)*
   - (a) Sodium meta bisulphate (b) Sodium benzoate
   - (c) Potassium meta bisulphate
   - (d) None of these
119. 'Buck eye rot' is a disease of *(Ph.D. VGS, 2005)*
   - (a) Sweet potato (b) Pea
   - (c) Tomato (d) Chilli
120. Leaf curl of tomato is spread by *(Ph.D. VGS, 2005)*
   - (a) White fly (b) Jassids
   - (c) Fungi (d) Leaf hopper
121. Pomato was produced by *(Ph.D. VGS, 2005)*
   - (a) Cytoplasmic fusion (b) Protoplasmic fusion
   - (c) Both (d) None of these
122. 'Sioux' is the variety of *(Ph.D. VGS, 2005)*
   - (a) Capsicum (b) Tomato
   - (c) Potato (d) Brinjal
123. Tip top is a variety of *(UP PCS (Pre), 2006)*
   - (a) Chilli (b) Tomato
   - (c) Brinjal (d) Capsicum
124. Stomp is used for weed control in brinjal at the rate of *(UP PCS (Pre), 2006)*
   - (a) 0.5 kg a.i./ha (b) 1.0 kg a.i./ha
   - (c) 1.5 kg a.i./ha (d) 2.0 kg a.i./ha
125. The most suitable time of transplanting sweet pepper is *(ARS, 2003)*
   - (a) March-April (b) Mid-February
   - (c) April-May (d) None of these

ANSWERS

1. (b)	2. (c)	3. (b)	4. (c)	5. (c)
6. (c)	7. (a)	8. (c)	9. (b)	10. (d)
11. (a)	12. (a)	13. (a)	14. (d)	15. (d)
16. (b)	17. (a)	18. (a)	19. (c)	20. (c)
21. (d)	22. (a)	23. (c)	24. (a)	25. (c)
26. (b)	27. (c)	28. (b)	29. (c)	30. (d)
31. (c)	32. (b)	33. (b)	34. (a)	35. (d)

36.	(a)	37.	(d)	38.	(a)	39.	(b)	40.	(c)
41.	(a)	42.	(b)	43.	(a)	44.	(c)	45.	(c)
46.	(a)	47.	(a)	48.	(b)	49.	(a)	50.	(c)
51.	(a)	52.	(c)	53.	(c)	54.	(b)	55.	(c)
56.	(d)	57.	(a)	58.	(b)	59.	(c)	60.	(a)
61.	(a)	62.	(b)	63.	(c)	64.	(b)	65.	(c)
66.	(d)	67.	(b)	68.	(b)	69.	(b)	70.	(a)
71.	(c)	72.	(d)	73.	(a)	74.	(b)	75.	(a)
76.	(d)	77.	(a)	78.	(c)	79.	(b)	80.	(a)
81.	(a)	82.	(b)	83.	(b)	84.	(a)	85.	(a)
86.	(a)	87.	(d)	88.	(d)	89.	(c)	90.	(c)
91.	(a)	92.	(c)	93.	(c)	94.	(a)	95.	(a)
96.	(d)	97.	(a)	98.	(b)	99.	(c)	100.	(a)
101.	(b)	102.	(c)	103.	(b)	104.	(d)	105.	(c)
106.	(c)	107.	(d)	108.	(d)	109.	(a)	110.	(c)
111.	(b)	112.	(b)	113.	(a)	114.	(b)	115.	(a)
116.	(a)	117.	(a)	118.	(b)	119.	(c)	120.	(a)
121.	(a)	122.	(b)	123.	(b)	124.	(b)	125.	(b)

CHAPTER 10

PERENNIAL VEGETABLES

This is an important group of vegetables which does not require planting every year.

10.1 Drumstick

- Drumstick belongs to family *Moringaceae*.
- Drumstick is cultivated for its leaves, flower buds and tender fruits, all of which are used as nutritious vegetable.
- Drumstick seeds contain an oil called bean or behen oil.
- The flowers of drumstick are white or creamy white.
- Drumstick is a tropical plant and grows well in the plains.
- Jaffna is a variety of drumstick.
- Perennial types of drumstick are propagated by limb cutting.
- Annual types of drumstick are propagated by seeds.

10.2 Horse radish

- The horse radish is a perennial crop grown for its pigmented roots.
- The pungency in horse radish is due to the presence of an allyl-isothiocyanate and butyl-thiocyanate. The combination of these two produces glucoside sinigrin.
- Horse radish is propagated by root cuttings. It is also propagated by crowns.

10.3 Breadfruit

- The breadfruit (*Artocarpus altilis* Postperg) is native of Malaysia.
- The fruits of the seedless breadfruit contain a high amount of carbohydrates (27.98%). It is rich in calcium.
- Yellow Heart is a variety of breadfruit.
- One tree of breadfruit bears 50-100 fruits weighing 25-50 kg annually.

10.4 Curry-leaf

- Curry-leaf is an important perennial tree vegetable-cum-spice crop of India.

- Ksenigin and Murayin are crystalline glucoside present in curry leaves and flowers, respectively.
- Suwasini is a variety of curry leaf.
- Curry-leaf is propagated through seeds.
- Polyembryony has been reported in curry-leaf.
- Fruit of curry-leaf contains two seeds per fruit and such seeds are separated by a thin papery seed coat.

10.5 Asparagus

- Asparagus is a perennial dioecious herb.
- The tender shoots called 'spears' are used as vegetables.
- Asparagus (*Asparagus officinalis* L.) belongs to family *Liliaceace*.
- Female flowers of asparagus plants produce berries with one to three seeds per fruit.
- The male flowers of asparagus are yellowish-green in colour.
- Blanching is practised to blanch the young spears and get 'white asparagus' for canning.

The objective type questions on perennial Vegetables are given below :

1. Asparagus is a
   (a) Perennial herb (b) Dioecious herb
   (c) Both (d) None of these
2. Edible part of asparagus is
   (a) Tender roots (b) Tender shoots
   (c) Seed (d) None of these
3. Asparagus tender shoots are called
   (a) Spears (b) Spathe
   (c) Both (d) None of these
4. Tender shoots of asparagus contain
   (a) Saponine (b) Asparagine
   (c) Both (d) None of these
5. *Asparagus officinalis* L. is the botanical name of
   (a) Rhubarb (b) Asparagus
   (c) Basella (d) None of these
6. *Asparagus officinalis* L. belongs to the family
   (a) *Leguminaceae* (b) *Liliaceace*
   (c) *Amaryllidaceae* (d) None of these
7. Female flowers of an asparagus plants produce berries with seeds
   (a) 1-3 (b) 5-10
   (c) 10-15 (d) 15-20

8. The colour of male flowers of asparagus is
   (a) Yellow (b) Green
   (c) Yellowish-green (d) None of these
9. Asparagus is propagated through
   (a) Seed (b) Crown
   (c) Both (d) None of these
10. Blanching is practised to blanch the young spears and get 'white asparagus' for
    (a) Vegetable purpose (b) Canning purpose
    (c) Both (d) None of these
11. Family of sea kale is
    (a) *Leguminaceae* (b) *Cruciferae*
    (c) *Solanaceae* (d) None of these
12. Edible portion of sea kale is
    (a) Young leaves (b) Shoots
    (c) Both (d) None of these
13. Sea Kale is a
    (a) Hardy perennial (b) Soft perennial
    (c) Hardy annual (d) None of these
14. Edible portion of rhubarb is
    (a) Leaf stalk (b) Petiole
    (c) Both (d) None of these
15. Ruby Valentine, Sunrise, Straw Berry and Cherry Red are the improved varieties of
    (a) Rhubarb (b) Sea kale
    (c) Asparagus (d) None of these
16. Victoria and Linnaeus are the oldest varieties of
    (a) Rhubarb (b) Sea kale
    (c) Asparagus (d) None of these
17. Rhubarb is a
    (a) Cold tolerant plant (b) Tropical plant
    (c) Sub tropical plant (d) None of these
18. The first harvesting in Rhubarb is started from the year
    (a) First (b) Second
    (c) Third (d) None of these
19. Rhubarb is propagated by
    (a) Root stock (b) Seed
    (c) Corms (d) None of these
20. Rhubarb is tolerant to
    (a) Drought (b) Water lodging
    (c) Both (d) None of these

21. Rhubarb is tolerant to
    (a) Acidity (b) Alkalinity
    (c) Both (d) None of these
22. *Moringa oleifera* Lam is the botanical name of
    (a) Drumstick or morning tree
    (b) Horse radish tree
    (c) Radish tree or west Indian bean
    (d) All the above
23. Drumstick belongs to the family
    (a) *Cruciferae* (b) *Compositae*
    (c) *Moringaceae* (d) None of these
24. Edible portion of drumstick is
    (a) Leaves (b) Floral buds
    (c) Tender fruits (d) All the above
25. Drumstick Seed contains
    (a) Oil (b) Protein
    (c) Both (d) None of these
26. Drumstick seed oil is called
    (a) Stick oil (b) Behen oil
    (c) Both (d) None of these
27. The flowers of drumstick are
    (a) White (b) Creamy white
    (c) Both (d) None of these
28. Drumstick is a
    (a) Tropical plant (b) Temperate plant
    (c) Sub-tropical plant (d) None of these
29. 'Jaffna' is a variety of
    (a) Moringa (b) Water leaf
    (c) Horse radish (d) None of these
30. Perennial type of drumstick is propagated through
    (a) Seeds (b) Limb cuttings
    (c) Both (d) None of these
31. Annual type of drumstick is propagated through
    (a) Seeds (b) Limb cutting
    (c) Both (d) None of these
32. Horse radish is a
    (a) Perennial crop (b) Annual crop
    (c) Both (d) None of these
33. Edible part of horse radish is
    (a) Pigmented roots (b) Shoots
    (c) Both (d) None of these

34. Pungency in horse radish is due to
    (a) Allyl-isothiocyanate (b) Butyl-thiocyanate
    (c) Glucoside sinigrin (d) All the above
35. Horse radish is propagated by
    (a) Root cuttings (b) Crowns
    (c) Both (d) None of these
36. *Artocarpus altilis* is the botanical name of
    (a) Horse Radish (b) Bread fruit
    (c) Drum stick (d) None of these
37. Breadfruit contains high amount of
    (a) Carbohydrate (b) Protein
    (c) Fat (d) None of these
38. Breadfruit contains carbohydrate
    (a) 10.2% (b) 15.50%
    (c) 25.7% (d) 27.98%
39. Breadfruit contains
    (a) One seed (b) Two seeds
    (c) Three seeds (d) None of these
40. 'Yellow Heart' is a variety of
    (a) Horse radish (b) Breadfruit
    (c) Drumstick (d) None of these
41. Breadfruit is propagated through
    (a) Root cuttings (b) Stem cuttings
    (c) Seed (d) None of these
42. Seedlessness in breadfruit is due to
    (a) Parthenocarpy (b) Stimulative parthenocarpy
    (c) Both (d) None of these
43. One tree of breadfruit bears
    (a) 1-10 Fruits (b) 10-50 Fruits
    (c) 50-100 Fruits (d) None of these
44. Annual yield per plant of breadfruit is
    (a) 10-20 kg (b) 20-25 kg
    (c) 25-50 kg (d) None of these
45. PKM-1 and PKM-2 are the improved varieties of
    (a) Breadfruit (b) Drumstick
    (c) Asparagus (d) None of these
46. Fruit weight of drumstick is
    (a) 120 g (b) 140 g
    (c) 150 g (d) 230 g

47. Native of Chekurmanis/Madura Keera is
(a) China (b) India
(c) Africa (d) None of these

48. Chekurmanis is also known as
(a) Vegetable of 10th Century
(b) Vegetable of 20th Century
(c) Vegetable of 21st Century (d) None of these

49. Chekurmanis is commonly known as
(a) Multivitamin greens (b) Multi protein greens
(c) Both (d) None of these

50. *Sauropus androgynus* is the botanical name of
(a) Chekurmanis (b) Water-leaf
(c) Basella (d) None of these

51. Chekurmanis is propagated through
(a) Root cuttings (b) Stem cuttings
(c) Seeds (d) None of these

52. Edible portion of Chekurmanis is
(a) Leaves (b) Tender shoots
(c) Both (d) None of these

53. Leaves and tender shoots of Chekurmanis are used for
(a) Leafy vegetables (b) Salad
(c) Both (d) None of these

54. *Talimum triangulare* is the botanical name of
(a) Ceylon spinach or water-leaf (b) Chekurmanis
(c) Basella (d) None of these

55. Native of ceylon spinach is
(a) India (b) Brazil
(c) U.S.A. (d) None of these

56. Mucilaginous leafy vegetable is
(a) Ceylon Spinach or Water-leaf (b) Chekurmanis
(c) Both (d) None of these

57. Shade loving leafy vegetable is
(a) Ceylon Spinach or Water-leaf (b) Chekurmanis
(c) Both (d) None of these

58. Water-leaf is a
(a) Long-day plant (b) Short-day plant
(c) Day-neutral plant (d) None of these

59. Water-leaf is propagated through
(a) Seeds (b) Stem cuttings
(c) Both (d) None of these

60. Basella is called
    (a) Indian spinach (b) Malabar spinach
    (c) Both (d) None of these
61. Native of Basella is
    (a) South Asia (b) South Africa
    (c) India (d) None of these
62. Edible portion of Basella is
    (a) Leaves (b) Petiole
    (c) Tender stem (d) All the above
63. Basella is propagated through
    (a) Seeds (b) Stem cuttings
    (c) Both (d) None of these
64. Drumstick leaves are the richest source of
    (a) Fat (b) Vitamin C
    (c) Both (d) None of these
65. Edible portion of artichoke is *(ARS, 2003)*
    (a) Root (b) Flower bud
    (c) Fruit (d) Leaf

ANSWERS

1. (c)	2. (b)	3. (a)	4. (b)	5. (b)
6. (b)	7. (a)	8. (c)	9. (c)	10. (b)
11. (b)	12. (c)	13. (a)	14. (c)	15. (a)
16. (a)	17. (a)	18. (b)	19. (c)	20. (a)
21. (a)	22. (d)	23. (c)	24. (d)	25. (c)
26. (b)	27. (c)	28. (b)	29. (a)	30. (b)
31. (a)	32. (a)	33. (a)	34. (d)	35. (c)
36. (b)	37. (a)	38. (d)	39. (d)	40. (b)
41. (b)	42. (b)	43. (c)	44. (b)	45. (b)
46. (d)	47. (b)	48. (c)	49. (a)	50. (a)
51. (b)	52. (c)	53. (c)	54. (a)	55. (b)
56. (a)	57. (a)	58. (b)	59. (c)	60. (c)
61. (a)	62. (a)	63. (c)	64. (c)	65. (b)

CHAPTER 11

SPICES AND CONDIMENTS

- India is known as home of spices.
- Largest area under spices – Rajasthan (0.76 m, ha).
- Highest production of spices- Andhra Pradesh (0.88 m, tonnes).
- Highest productivity under spices – Andhra Pradesh (2.5 t/ha).
- India's share in world trade of spices is around- 18%.
- Spice oil is obtained by steam distillation method.
- Black pepper and turmeric constitute more than 50% of total export earnings among spices.
- Total area under spices in India- 3.22 m,ha.
- Total production of spices in India- 3.76 m, tonnes.
- Total productivity of spices in India- 1.1 t/ha.
- Leading spices producing states- A.P. (.88 mt) > Rajasthan (.61 mt) > Gujarat (50 mt).

11.1 Production Share of different states -

1. Kerala { Black pepper 94% Production; Cardamon 60% Production; Ginger 25% Production }
2. A.P. { Turmeric; Chilli }
3. Rajasthan { Fenugreek-(80% Area); Fennel; Coriander }
4. Gujarat Cumin
5. USA-major importer and consumer of oleoresins (50%).

11.1.1 State-wise area production and productivity of major spices in India

State	Area in (000'ha)	Production in (000'mt)	Productivity in (mt/ha)
Andhra Pradesh	354.90	887.60	2.5
Rajasthan	762.60	616.40	0.80
Gujarat	473.50	505.30	1.10
Karnataka	284.70	315.70	1.10
Tamil Nadu	159.30	250.40	1.60
Orissa	253.80	224.00	0.90
Madhya Pradesh	171.50	152.90	0.90
Kerala	288.10	151.60	0.50
Maharashtra	121.30	104.30	0.90
Uttar Pradesh	42.40	100.60	2.40
West Bengal	86.40	95.00	1.10
Others	225.40	362.30	1.10
Total	3223.90	3766.10	1.16

11.1.2 Crop-wise area, production and productivity of major spice crops

Spice crop	Area (000'ha)	Production (000'mt)	Productivity (mt/ha)
Chilli I	881.30 I	1113.1 I	1.30
Turmeric	163.00	552.00 II	3.40 III
Mustard seed III	467.8 III	410.50 III	0.90
Garlic	88.10	367.60	4.20 I
Coriander seed	433.40	318.70	0.70
Ginger	84.60	317.80	3.80 II
Cumin seed II	526.60 II	206.40	0.40
Tamarind	60.80	184.40	3.00
Fenugreek seed	115.60	136.60	1.20
Pepper	216.50	79.10	0.40
Fennel seed	39.90	38.50	1.00
Cardamom	102.70	17.30	0.20
Other	39.70	22.70	-
Total	3220.00	3765.00	1.17

11.1.3 Production share of major spices in India

S.No.	Crop	Production share in %
1.	Chilli	29.60
2.	Turmeric	14.70
3.	Mustard seed	10.90
4.	Garlic	9.80
5.	Coriander seed	8.50
6.	Ginger	8.40
7.	Cumin seed	5.50
8.	Tamarind	4.90
9.	Fenugreek seed	3.60
10.	Pepper	2.10
11.	Fennel seed	1.00
12.	Cardamom	0.50
13.	Others	0.60

- National Research Center for seed spices- Tabiji, Rajasthan
- AICSIP- All India Coordinated Spices Improvement Project
- ITC- International Trade Center
- IWGS- International Working Group on Spices (1979)
- SEC- spices Enquiry Committee (1951-53)
- Indian spices and cashew nut committee (1961)
- IISR- Indian Institute of Spices Research,Calicut (Kerala)
- National Research center for spices, Calicut (Kerala)
- Directorate on seed and spices, Mahu (Varanasi)
- ISO-International Organization for standardization
- Spices crops are largely grown in tropics between 20°N and 20°S
- CAC – Codex Alimentarius Commission (1962)
- Agmark – Agriculture marketing (1937)
- In India total species grown - 52
- Total species in world according to International standard organization – 107
- Major spices- 5(Black pepper, Cardamom, Chilli, Ginger, Turmeric)
- Miner spices- 50

11.2 Black Pepper

Botanical name – *Piper nigrum*

Family – Piperaceae

Chromosome no. – 2n = 128

Origin – Indo-Burma

- King of Spices
- Largest producer country in the world- Vietnam
- India accounts for 54% of total area and 26% of total production in the world.
- Fruiting branches – Plagiotropes
- Top shoots – Orthotropes
- Hanging shoots – Geotropes
- Propagation – By shoot cutting
- Average yield – 273 kg/ha
- Harvesting time – Nov. – Feb. in plains
Jan. – March in hills

Kerala accounts – 94% of total production

Varieties - Karimunda
Sreekara
Poornami
Paniyur 2,4,6
Subhakara
Panchami

- Hybrids- Panniyur-1
Panniyur-3
- Self sterile variety – Balankotta, Kalluvalli
- Tripod – Stand for harvesting
- North America – Major importer of Indian pepper
- Plant part used – Fruit and Seed
- Important Chemical content – Piperine
- Essential oil - 1-3 %

11.3 Small Cardamom

Botanical name – *Eletaria Cardamomum*

Family – Zingeberaceae

Chromosome no. – 2n = 24

Origin – India

- Queen of spices
- Propagation by rhizomes
- Harvesting time – October- November
- II Important national spice

- Inflorescence – Panicle
- Yield- 500 kg (5 q/hac.)
- Summer showers are essential for flowering

Varieties- Mysore type
Malabar type
PV-2
Mudigree-1
Vazukha type- cross between Mysore x Malabar.

- Plant part used - Fruit and Seed
- Chemical content - Cineol
- Essential oil - 1-2 %

11.4 Large Cardamom

Botanical name – *Amomum subulatum*
Family – Zingeberaceae
Chromosome no. – 2n = 24
Origin – India

- It is shade loving plant
- Propagation - Suckers

Varieties- Bebo
Golsey
Ramla
Ramsey

- Chemical content - Cineol
- Essential Oil - 1-2 %

11.5 Ginger

Botanical name – *Zingiber officinalis*
Family – Zingiberaceae
Chromosome no. – 2n = 22
Origin – South-East Asia

- India accounts 50% of total world production and 50% in world export.
- Kerala is leading producer of ginger (60% area & 25% production)
- Propagation by Rhizomebits of 15-20 g @ 12-18 q/ha
- Average yield – 150—300 q/ha.
- Varieties –
  1. Riode Janerio (Introduction)
  2. Surbhi (Mutant)
  3. Suphrabha
  4. Suruchi
  5. Wynad manantody – Contribute 50% of total production
  6. Himgiri – tolerant to rhizome rot

- Cochin ginger/Indian ginger best in the world.
- Plant part used - Rhizome
- Chemical content - Zingeberene/Gengerol
- Essential oil - 1.9 - 2.2%
- Seed rate - 15-16 q/ha
- Sowing time - Pre-monsoon
- Transplanting space - 30 X 20 cm
- Crop harvest time - 6 – 8 months
- Yield - 15-30 t/ha
- Dry Zinger (yield) - 20-25% of total yield.

11.6 Turmeric

Botanical name – *Curcuma longa* L.

Family – Zingiberaceae

Chromosome no. – 2n = 32

Origin – South-East Asia

- India accounts – 76% of total world production
- Andhra Pradesh leading turmeric producer state
- Propagation – by rhizome
- Seed rate – 25 q/ha
- Average yield – 25-30 t/ha
- Trade type – Allepy finger turmeric, famous for its colour in trade
- Varieties –

Rajendra sonia	Sudarshan
Pratibha	Sugandham
Prabha	Saguna
Kranti	Suroma
Krishna	Suvarna
Rashmi	Allepy turmeric – Best in the world
Roma	Azad Haldi-1
Sobha	

- Turmeric is antioxidant due to phenolic character of curcumin
- Part used – Rhizome
- Chemical content – Curcumine, Essential Oil = 0.24%
- Crop harvest time – 7-9 month
- Transplanting space – 30 X 20 cm
- Rizome weight for sowing – 20-40 g
- Sowing time – June- July (Pre-monsoon)

11.7 Clove (Single Seeded Drupe)

Botanical name – *Syzygium aromaticum*

Family – Myrtaceae

Chromosome no. – 2n = 44

Origin – Indonesia

- Indonesia account for 68% of world production
- About 11,000-15,000 dried cloves makes 1 kg

Varieties –

Pevang
Lenzipore
Amulya

- Alternate bearing habit
- IInd most important spice next to black pepper in the world
- Part used - Unopened flower bud
- Chemical content - Eugenol
- Propagated by seeds

11.8 Nutmeg

Botanical name – *Myristica fragrans*

Family – Myrsticaceae

Origin – Indonesia

- Nutmeg – dried kernel of seed
- Mace – dried aril surrounding the seed
- Harvesting time – June-August
- Dioecious tree
- Average weight of individual nutmeg – 60 g
- Seed weight 6-7g, Mace- 3-4 g
- Pericarp is used for making Jam, Jelly, pickles
- Unproductive male is converted into female by top working
- Pericarp- rich source of pectin (12-14%)
- Volatile oil has weedicidal property
- Part used - Seed, Aril
- Chemical content - Terbein

11.9 Cinnamon

Botanical name – *Cinnamomum verum*

Family – Lauraceae

Chromosome no. – 2n = 24

Origin – Sri Lanka

- Tejpat (Indian cassia) – Cinnamon tamala
- Oldest known spice
- Commercially propagated method- Seed
- Harvesting time – Sep.-Nov.
- Varieties –

Navashree	Konkan Tej
Nethyashree	Yercaud- 1

- Sri Lanka largest producer of cinnamon bark
- Part used - Bark
- Chemical content - Cinamaldehyde
- Essential oil - 0.5 – 2%

11.10 Fenugreek

Botanical name – (i) Common methi - *Trigonella foenum graecum*
(ii) Kasuri methi - *Trigonella corniculatum*

Family – Leguminaceae/Fabaceae

Chromosome no. – 2n = 16

Origin – Europe

- Seed contain diosgenin – Which is used to prepare contraceptive pills
- More than 80% area and production contributed by Rajasthan
- Sowing time – Last week of October
- Seed Rate – 25 kg/ha
- Average yield 10-11 q/ha
- It is third largest seed spice in India after coriander and cumin
- Variety-

Rajendra Kranti	Pusa Early Bunching
Hissar sonali	Gujarat methi-1
Azad methi-1	LAM- Selection-1
Kasuri methi	

- Fenugreek is both tropical and temprate crop.
- India is major producer and exporter of fenugreek seed
- Harvesting time- February – March (100-120 days crop)
- Bitter taste in fenugreek – "Trigonella and Choline" (7%)
- Part used - Green leaves and Seed
- Chemical content - diosgenin

11.11 Coriander

Botanical name – *Coriandrum sativum*

Family – Apiaceae/Umbelliferae

Chromosome no. – 2n = 22

Origin – Mediterranean region

- Rajasthan leads in area and production
- Drying is done to maintain moisture level below 13% in seeds.
- Seed rate 25-30 kg/ha (Rain fed), 12-15 kg/ha (irrigated) 40-42 kg/ha (leaves)
- yield 18-20 q/ha (Seed)
- Spacing - 30 X 15 or 30 X 10 cm
- Varieties –

Sindhu	Sadhna
Swathi	Rajendra Swathi
Merrocan	Pant Haritma
Karan-Small seeded variety	Azad Dhania-1

- Sowing time : October-November
- Leaves contain - Protein 3.3%
  Fat 0.6%
  Carbohydrate 6.5%
  Mineral 1.7%
- Chemical quality - 0.1-2.7% essential oil
  10-20% Fait oil
- Leading seed spice grown in India
- Most heavily consumed herb plant
- Fruits- Schizocarp
- Part used - Leaves, Seed
- Chemical content - “Linalool & Borneol”

11.12 Cumin (Zeera)

Botanical name – *Cuminum cyminum* L.

Family – Apiaceae/Umbelliferae

Chromosome no. – 2n = 14

Origin – Mediterranean region

- Area and Production - Ist position Gujarat
- Best sowing time - November
- Seed rate - 12-15 kg/ha
- Spacing - 30 X 15 Cm
- Yield - 7-8 q/ha
- Varieties –

RZ-19	GJ- 1,3,4
GC-1	
Vijapur-5	
RZ- 209	

- Fruits rich in thymol
- Iran- main cumin exporter
- Storage – 9% Humidity
- Part used - Fruit
- Chemical content - 'Cuminol'
- Essential oil - 2.5-3.5%
- Flower colour - White or Pink
- Fruits - Schizocarpic

11.13 Fennel (Sauf)

Botanical name – *Foeniculum vulgare*
Family – Apiaceae/ Umbelliferae
Chromosome no. – 2n = 22
Origin – Mediterranean region

- Best sowing time- September-October
- Seed Rate- 10-12 kg/ha
- Yield- 9-10 q/ha
- Varieties :

  RF – 101 Azad Sauf-1
  RF – 125
  RF – 35

- Spacing – 45 X 30 cm
- Storage moisture – 8-9%
- Part used - Fruit
- Chemical content - 'Limone'
- Essential oil - 1-3%
- Chemical content response for sweet taste 'Anithole' (50-60%)
- Flower colour-yellow

11.14 Aniseed (Vilayati sauf)

Botanical name – *Pempinella anisum*
Family – Apiaceae
Chromosome no. – 2n = 22
Origin – Mediterranean region

- Part used - Fruits
- Chemical content - 'Anithol'

11.15 Allspice (Double Seeded Berry)

Botanical name – *Pimenta dioca*
Family – Myrtaceae
Origin – West Indies

- Jamaica – Major producer of allspice
- Berries rich Vitamin – A (144 SIO)
- It contains flavour of Cinnamon, Clove and Nutmeg
- Functionally, it is dioecious
- Edible part-dried Immature fruits (Berries)
- Cluster shows (+) ve response to flowering and fruiting in allspice
- Part used - Fruit seed
- Chemical content - Phenol

11.16 Curry Leaf

Botanical name – *Murrya Koenigi*

Family – Rutaceae

Chromosome no. – 2n = 18

Crystaline glucoside 'Ksenigin' obtained from leaves, 'Muragin' obtained from flowers.

- Self pollinated crop
- Rich in Ca
- Variety - Suwasani
- Propagation by seeds
- Part used - Leaves
- Chemical content - Koenigine

11.17 Kokum

Botanical name -- *Garcinia indica*

Family – Guttiferae

Chromosome no. – 2n = 24

Origin – India

- Dioecious in nature
- Variety - Konkan Amrit
- Kokum contain - Hydroxy citric acid
- Part used - Fruit
- Chemical content - β-hydroxy acetic acid

11.18 Vanilla

Botanical name – *Vanilla planifalia*

Family – Orchidaceae

Origin – America

- Madagasy Republic (70-80%) is leading producer in world.
- Natural pollination is impossible due to presence of rostellum which prevents contact of anther cap and stigma.
- Propagation – Stem cutting

- Oil is called balsam of vanilla (Vanillin) – 2.3- 2.9% in beans
- Vanilla is developed due to action of enzymes (β-glycoside)
- Vanilla is extracted by hydro alcoholic extraction
- Caumarin –Cheep substitute for vanilla
- Tender, green pods are harvested for use
- Part used – Fruit
- Chemical content – Vanillin

11.19 Saffron (Kesar)

Botanical name – *Crocus satiyus*
Family – Iridaceae
Origin – India

- Fennel shaped stigma is used as spice
- In India – cultivated only in Kashmir
- Most expensive saffron – Shahi saffron
- About 1.5 millions flower on drying give 1 kg of saffron
- Propagated by corms
- Yield – 5 kg dried saffron/hac
- Part used – Stigma
- Chemical content – Cicrocrocin
- Essential oil – 0.5-1.0 %

11.20 Asafoetida (Heeng)

Botanical name – *Ferulla foetida*
Family – Apiaceae

- Stage of tapping – Foliace turn green to yellow
- Hadda – most priced and strongest trade
- Ayurvedic medicine - 'Hingachakta' is prepared from asafoetida
- Part used - Olearesin Gum
- Chemical Content - Ferumine
- Essential Oil - 10-17%

11.21 Ajowain (Caraway)

Botanical name – *Trachyspermum ammi* Linn
Family – Umbelliferae
Origin – South-East Asia

- Part used - Fruits (Seeds)
- Producing state - U.P.
- Seed rate - 3-4 kg/ha, Space 30 X 15 cm
- Sowing time - Oct- Nov

11.22 Dill (Soya)

Botanical name – *Anethum graveolens*
Family – Apiaceae/Umbelliferae
Origin – Eurasia

- Part used - Fruit/seed
- Chemical component – Carvone
- Essential oil - 1.5 – 4.0%
- Seed use for making gripe water
- Varieties :
  NRCSS – A.D.-1
  NRCSS – A.D.-2
- Sowing time - August – Early , October – Rabi
- Flower colour - Yellow
- Seed rate - 2-5 kg/ha
- Space - 45 X 20 cm

11.23 Nigella/ Kallogi / Mangral

Botanical name – *Nigella sativa* L.
Family – Ramiculaceae
Origin – South-East Asia & America

- Seed rate – 10 kg/ha, Yield-10-15 q/ha
- Part used – Fruit/seed (Capsulated black colour)
- Space - 30 X 15 cm
- Sowing time - Oct. – Nov.
- Varieties - Azad Kalongi-1

The objective type questions on spices and condiments are given below :

1. Which country is known as home of spices ?
   (a) China (b) India
   (c) Pakistan (d) U.S.A.

2. Total area under spices in India is *(Ph.D. VGS, 2005)*
   (a) 2.76 m ha (b) 3.22 m ha
   (c) 4.86 m ha (d) None of these

3. Total production of spices in India is
   (a) 3.76 mt (b) 3.86 mt
   (c) 4.76 mt (d) None of these

4. Which state has largest area under spices ?
   (a) Rajasthan (b) Andhra Pradesh
   (c) Gujarat (d) None of these

5. Which state has highest production of spices ?

(Ph.D. VGS, 2004)

(a) Rajasthan (b) Andhra Pradesh
(c) Gujarat (d) None of these

6. Which state has highest productivity of spices ?

(a) Andhra Pradesh (b) Gujarat
(c) Rajasthan (d) None of these

7. India's share in world trade of spices is around

(Ph.D. VGS, 2004)

(a) 16% (b) 18%
(c) 20% (d) None of these

8. Spice oil is obtained by the process of

(a) Steam distillation (b) Water distillation
(c) Both (d) None of these

9. Which country is major importer and consumer of oleoresin in the world ?

(a) India (b) China
(c) USA (d) None of these

10. Leading state in production of black pepper, cardamom and ginger in the world is

(a) Kerala (b) Andhra Pradesh
(c) Rajasthan (d) Gujarat

11. The state having highest production of turmeric and chilli is

(a) Andhra Pradesh (b) Kerala
(c) Karnataka (d) Rajasthan

12. The state having highest production of fenugreek, fennel and coriander is

(a) Andhra Pradesh (b) Rajasthan
(c) Kerala (d) Gujarat

13. The state having highest production of cumin is

(a) Gujarat (b) Rajasthan
(c) Andhra Pradesh (d) Karnataka

14. National Research Centre for seed spices is situated in

(a) Andhra Pradesh (b) Rajasthan
(c) Gujarat (d) Kerala

15. International working group on spices was established in

(ARS, 2004)

(a) 1979 (b) 1980
(c) 1981 (d) 1982

16. Indian Institute of spices research is situated in

(a) Karnataka (b) Kerala
(c) Andhra Pradesh (d) Tamil Nadu

17. National Research Centre for spices is situated in
    (a) Kerala (b) Karnataka
    (c) Tamil Nadu (d) Andhra Pradesh
18. Directorate on seed spices is situated in *(MP. PCS, 2003)*
    (a) Lucknow (b) Varanasi
    (c) New Delhi (d) Bangalore
19. Spice crops are largely grown in tropics between *(ARS, 2004)*
    (a) 20°N and 20°S (b) 20°W and 20°E
    (c) 20°N and 20°W (d) 20°S and 20°E
20. Total number of spices grown in India are *(Ph.D. VGS, 2005)*
    (a) 52 (b) 48
    (c) 59 (d) 62
21. Botanical name of black pepper is
    (a) *Piper nigrum* (b) *Eletaria*
    (c) *Amomum subulatum* (d) *Capsicum annum*
22. Black pepper belongs to the family
    (a) Solanaceae (b) Piperaceae
    (c) Zingiberaceae (d) Myrtaceae
23. Somatic chromosome number of black pepper is *(ARS, 2004)*
    (a) 110 (b) 128
    (c) 136 (d) 144
24. Largest black pepper producing country in the world is
    (a) Indonesia (b) USA
    (c) Vietnam (d) India
25. Propagation of black pepper is done through
    (a) Shoot cuttings (b) Root cuttings
    (c) Leaf cuttings (d) Seed cutting
26. Average yield of black pepper per hectare is *(Ph.D. VGS, 2005)*
    (a) 273 kg (b) 290 kg
    (c) 283 kg (d) 260 kg
27. Harvesting of black pepper in plains is done in the month of
    (a) January – March (b) November – February
    (c) June – August (d) April – July
28. Harvesting of black pepper in hills is done in the month of
    (a) November – February (b) January – March
    (c) April – June (d) July – September
29. Origin of black pepper is
    (a) Indo-Burma (b) China
    (c) USA (d) Pakistan

30. Karimunda, Sreekara, Poornami, Subhakara and Punchami are the improved varieties of
    (a) Chilli (b) Black pepper
    (c) Cardamom (d) Clove

31. Panniyur-1 and Panniyur-3 are the varieties of
    (a) Black pepper (b) Clove
    (c) Cardamom (d) Chilli

32. Balamkatta and Kalluvalli are the varieties of black pepper which are
    (a) Self fertile (b) Self sterile
    (c) Male sterile (d) None of these

33. Which part of black pepper is used ?
    (a) Fruit (b) Seed
    (c) Both (d) None of these

34. Which chemical is found in black pepper ?
    (a) Piperine (b) Oleoresin
    (c) Cineol (d) Linalool

35. Essential oil content in black pepper is *(M.Sc. Ag., 2004)*
    (a) 4-5% (b) 1-3%
    (c) 6-9% (d) 10-12%

36. Botanical name of small Cardamom is
    (a) *Eletaria cardamom* (b) *Piper nigrum*
    (c) *Amomum subulatum* (d) *Capsicum annum*

37. Small cardamom belongs to the family
    (a) *Zingiberaceae* (b) *Solanaceae*
    (c) *Piperceae* (d) *Myrtaceae*

38. Somatic chromosome number in small cardamom is
    (a) 22 (b) 24
    (c) 36 (d) 18

39. Origin of small cardamom is
    (a) China (b) India
    (c) USA (d) Afghanistan

40. Queen of spices is called
    (a) Small cardamom (b) Large cardamom
    (c) Clove (d) Black pepper

41. Propagation of small cardamom is done through
    (a) Suckers (b) Seeds
    (c) Rhizomes (d) Shoot cutting

42. Harvesting of small cardamom is done in the month of
    (a) August – September (b) October – November
    (c) December - January (d) February – March

43. Inflorescence of small cardamom is known as *(ARS, 2005)*
    (a) Recemose (b) Cymose
    (c) Panicle (d) None of these
44. Average yield of small cardamom per hectare is
    (a) 500 kg (b) 700 kg
    (c) 900 kg (d) None of these
45. Sunny showers are essential for flowering in
    (a) Large cardamom (b) Small cardamom
    (c) Black pepper (d) Clove
46. Mysore type, Malabar type, PV-2 and Mudigree-1 are the varieties of
    (a) Small cardamom (b) Clove
    (c) Black pepper (d) Large cardamom
47. Which chemical is found in small cardamom ?
    (a) Linalool (b) Oleoresin
    (c) Cineol (d) Gengerol
48. Essential oil content in small cardamom is
    (a) 1-2% (b) 3-4%
    (c) 5-6% (d) None of these
49. Botanical name of large cardamom is
    (a) *Amomum subulatum* (b) *Eletaria cardamom*
    (c) *Piper nigrsum* (d) *Syzygium aromaticum*
50. Large cardamom belongs to the family
    (a) Piperaceae (b) Zingiberaceae
    (c) Myrtaceae (d) Solanaceae
51. Somatic chromosome number in large cardamom is
    (a) 24 (b) 22
    (c) 26 (d) 28
52. Origin of large cardamom is
    (a) China (b) USA
    (c) India (d) Burma
53. Shade loving plant is
    (a) Large cardamom (b) Black pepper
    (c) Small cardamom (d) None of these
54. Propagation of large cardamom is done by
    (a) Rhizomes (b) Suckers
    (c) Seeds (d) Stem cuttings
55. Bebo, Golsey, Ramla and Ramsey are the varieties of
    (a) Large cardamom (b) Small cardamom
    (c) Ginger (d) Black pepper

56. The chemical found in large cardamom is *(Ph.D. VGS, 2005)*
    (a) Cineol (b) Linalool
    (c) Gengerol (d) Oleoresin
57. Essential oil content in large cardamom is
    (a) 3-4% (b) 1-2%
    (c) 5-7% (d) None of these
58. Botanical name of ginger is *(M.Sc. Ag., 2005)*
    (a) *Curcuma longa* (b) *Zingiber officinalis*
    (c) *Syzygium oromaticum* (d) *None of these*
59. Ginger belongs to the family
    (a) *Solanaceae* (b) *Myrtaceae*
    (c) *Zingiberaceae* (d) *Piperaceae*
60. Somatic chromosome number of ginger is
    (a) 22 (b) 24
    (c) 32 (d) 28
61. Origin of ginger is
    (a) South-West Asia (b) South-East Asia
    (c) USA (d) None of these
62. Leading producer state for ginger is *(MP. PCS, 2004)*
    (a) Andhra Pradesh (b) Kerala
    (c) Tamil Nadu (d) Karnataka
63. Ginger is propagated by
    (a) Suckers (b) Rhizomes
    (c) Bulbs (d) Seeds
64. Size of ginger rhizomes which is suitable for propagation is
    (a) 15-20 g (b) 25-40 g
    (c) 50-60 g (d) None of these
65. How much quantity of rhizomes is used for one hectare planting of ginger crop ?
    (a) 12-18 q (b) 20-25 q
    (c) 26-30 q (d) None of these
66. An average yield of ginger per hectare is *(Ph.D. VGS, 2005)*
    (a) 100-125 q (b) 150-300 q
    (c) 350-500 q (d) None of these
67. Rio de Janeiro variety of ginger has been developed through
    (a) Introduction (b) Pure line selection
    (c) Mass selection (d) None of these
68. Variety of ginger is
    (a) Suprabha (b) Surbhi
    (c) Suruchi (d) None of these

69. Rhizome rot tolerant variety of ginger is
    (a) Himgiri (b) Surabhi
    (c) Suruchi (d) None of these
70. Which part of ginger is used ?
    (a) Rhizomes (b) Suckers
    (c) Tubers (d) None of these
71. The chemical found in ginger is
    (a) Zingeberance (b) Oleoresin
    (c) Cineol (d) None of these
72. Essential oil content in ginger is
    (a) 2.9 – 3.4% (b) 1.9 – 2.2%
    (c) 3.2 – 4.5% (d) None of these
73. Transplanting space of ginger is
    (a) 30 x 20 cm (b) 30 x 30 cm
    (c) 30 x 40 cm (d) 30 x 50 cm
74. Harvesting time of ginger rhizomes is after
    (a) 4-5 months (b) 6-8 months
    (c) 9-11 months (d) None of these
75. Dry ginger yield per hectare from total yield is
    (a) 20-25% (b) 30-35%
    (c) 40-45% (d) None of these
76. Botanical name of turmeric is
    (a) *Zingiber officinalis* (b) *Curcuma longa*
    (c) *Piper nigrum* (d) *Capsicum annum*
77. Turmeric belongs to the family *(Ph.D. VGS, 2005)*
    (a) Myrtaceae (b) Solanaceae
    (c) Zingiberaceae (d) None of these
78. Somatic chromosome number in turmeric is *(Ph.D. VGS, 2005)*
    (a) 2n = 32 (b) 2n = 62
    (c) 2n = 63 (d) 2n = 74
79. Origin of turmeric is
    (a) India (b) China
    (c) South-East Asia (d) South-West Asia
80. Turmeric is propagated through
    (a) Suckers (b) Rhizomes
    (c) Bulbs (d) None of these
81. Per hectare seed rate of turmeric is
    (a) 20 q (b) 25 q
    (c) 30 q (d) 35 q

82. Per hectare average yield of turmeric is
(a) 150-200 q (b) 250-300 q
(c) 350-400 q (d) None of these

83. Allepy Finger turmeric is famous for trade due to *(Ph.D. VGS, 2005)*
(a) Colour (b) Quality
(c) Both (d) None of these

84. Sudarshan and Sugandham are the improved varieties of
(a) Ginger (b) Turmeric
(c) Clove (d) Garlic

85. Turmeric is antioxidant due to phenolic character of *(M.Sc. Ag., 2005)*
(a) Curcumin (b) Quersition
(c) Capsaicin (d) Cucurbitacin

86. Which part of turmeric is used ?
(a) Suckers (b) Rhizomes
(c) Bulbs (d) None of these

87. The chemical found in turmeric is
(a) Curcumin (b) Capsaicin
(c) Diallyl disulphide (d) Cuercitin

88. Harvesting time of turmeric after planting is
(a) 5-6 months (b) 7-9 months
(c) 10-12 months (d) 14-15 months

89. Transplanting space of turmeric is
(a) 30 x 20 cm (b) 30 x 30 cm
(c) 30 x 40 cm (d) 30 x 60 cm

90. Rhizome weight of turmeric for sowing is
(a) 10-12 g (b) 20-40 g
(c) 50-60 g (d) 70-80 g

91. Sowing time of turmeric is
(a) June-July (b) August – September
(c) February – March (d) April – May

92. Origin of clove is *(ARS, 2003)*
(a) India (b) China
(c) Indonesia (d) USA

93. Pevang, Lenzipore and Amulya are the improved varieties of
(a) Turmeric (b) Clove
(c) Small cardamom (d) Nutmeg

94. Alternate bearing habit is found in
(a) Clove (b) Nutmeg
(c) Cinnamom (d) Turmeric

95. Edible part of clove is
    (a) Opened flower bud (b) Unopened flower bud
    (c) Both (d) None of these

96. The chemical found in clove is
    (a) Linalool (b) Cineol
    (c) Eugenol (d) Terbein

97. Clove is propagated through
    (a) Seeds (b) Suckers
    (c) Bulbs (d) Rhizomes

98. Botanical name of nutmeg is *(ARS, 2005)*
    (a) *Linnamomum verum* (b) *Myristica fragrans*
    (c) *Syzygium aromaticum* (d) *Lumimum cyminum*

99. Nutmeg belongs to the family
    (a) *Lauraceae* (b) *Myrsticaceae*
    (c) *Myrtaceae* (d) *Solanaceae*

100. Dioecious tree is
    (a) Nutmeg (b) Clove
    (c) Cinnamon (d) Cumin

101. Average weight of individual nutmeg is
    (a) 20 g (b) 40 g
    (c) 60 g (d) 70 g

102. Which part of nutmeg is used for making Jam, Jelly and Pickles ?
    (a) Epicarp (b) Mesocarp
    (c) Pericarp (d) None of these

103. Unproductive male is converted into female by
    (a) Base working (b) Top working
    (c) Both (d) None of these

104. Pericarp of nutmeg is rich source of
    (a) Pectin (b) Protein
    (c) Fat (d) Carbohydrates

105. The percentage of pectin found in nutmeg is
    (a) 8-10% (b) 12-14%
    (c) 16-18% (d) 22-24%

106. Which part of nutmeg is used ?
    (a) Seed (b) Aril
    (c) Both (d) None of these

107. The chemical found in nutmeg is
    (a) Cineol (b) Linalool
    (c) Terbin (d) Eugenol

108. Botanical name of cinnamon is (ARS, 2003)
(a) *Cinnamomum verum* (b) *Trigonella corniculatum*
(c) *Myrstica fragrans* (d) *Cuminum cyminum*

109. Cinnamon belongs to the family
(a) Myrtaceae (b) Lauraceae
(c) Leguminaceae (d) Solanaceae

110. Somatic chromosome number in cinnamon is
(a) 24 (b) 22
(c) 26 (d) 28

111. Origin of cinnamon is
(a) India (b) China
(c) Sri Lanka (d) Pakistan

112. Cinnamon is propogated through
(a) Rhizomes (b) Bulbs
(c) Seeds (d) Stem cutting

113. Harvesting of cinnamon is done in the month of
(a) July-August (b) September – November
(c) December – January (d) February – April

114. Navashree, Nethyashree, Konkon Tej and Yarcaud-1 are the improved varieties of
(a) Nutmeg (b) Cinnamon
(c) Cumin (d) Coriander

115. The country producing largest cinnamon bark is
(a) Sri Lanka (b) India
(c) China (d) USA

116. Which part of cinnamon is used
(a) Seed (b) Bark
(c) Leaf (d) Stem

117. The chemical found in cinnamon is
(a) Eugenol (b) Cineol
(c) Cinamaldehyde (d) Gengerol

118. Essential oil content in cinnamon is (ARS, 1999)
(a) 0.1 – 0.4% (b) 0.5 – 2%
(c) 3-4% (d) 5-6%

119. Botanical name of Fenugreek is
(a) *Trigonella foenum graceum*
(b) *Trigonella cormiculatum*
(c) *Cinnamomum verum* (d) *Corindrum sativum*

120. Botanical name of Kasuri methi is
(a) *Trigonella foenum graecum*
(b) *Trignella corniculatum*
(c) *Cuminum cyminum* (d) *Foeniculum vulgare*

121. Fenugreek belongs to the family *(ARS, 1998)*
(a) *Solanaceae* (b) *Leguminosae*
(c) *Myrtaceae* (d) *Apiaceae*

122. Somatic chromosome number of fenugreek is
(a) 16 (b) 18
(c) 20 (d) 22

123. Origin of fenugreek is
(a) India (b) Europe
(c) Asia (d) China

124. The chemical found in fenugreek is
(a) Eugenol (b) Diosgenin
(c) Gengerol (d) None of these

125. Sowing time of fenugreek is
(a) Last week of October (b) First week of October
(c) Last week of November (d) First week of December

126. Per hectare seed rate of fenugreek is
(a) 20 kg (b) 25 kg
(c) 30 kg (d) 35 kg

127. Per hectare average yield of fenugreek is
(a) 7-9 q (b) 10-11 q
(c) 12-13 q (d) 14-15 q

128. Rajendra Kranti, Pusa Early Bunching, Azad Methi-1, Lam Selection-1, Kasuri Methi, Hisar Sonali and Gujarat Methi-1 are the improved varieties of
(a) Methi (b) Garlic
(c) Nutmeg (d) Cumin

129. Fenugreek is a crop of *(Ph.D. VGS, 2005)*
(a) Tropical region (b) Temperate region
(c) Both (d) None of these

130. The major producer and exporter of fenugreek seed is
(a) India (b) China
(c) Sri Lanka (d) Pakistan

131. Harvesting of fenugreek is done in the month of
(a) December-January (b) February-March
(c) April-May (d) August-September

132. Bitter taste in fenugreek is due to *(Ph.D. VGS, 2005)*
(a) Trigonella (b) Choline
(c) Both (d) None of these

133. Which part of fenugreek is used ?
(a) Green leaves (b) Seeds
(c) Both (d) None of these

134. Botanical name of coriander is
(a) *Coriandrum sativum* (b) *Cuminum cyminum*
(c) *Foeniculum vulgare* (d) *Garcinia indica*

135. Coriander belongs to the family
(a) *Apiaceae* (b) *Umbelliferae*
(c) Both (d) None of these

136. Somatic chromosome number in coriander is
(a) 22 (b) 24
(c) 26 (d) 28

137. Origin of coriander is
(a) Europe (b) Mediterranean region
(c) India (d) China

138. Drying is done in coriander seeds to maintain moisture level upto
(a) 16% (b) 14%
(c) 13% (d) None of these

139. Per hectare seed rate of coriander in rainfed conditions is
(a) 15-20 kg (b) 25-30 kg
(c) 35-40 kg (d) 45-50 kg

140. Per hectare seed rate of coriander in irrigated conditions is
(a) 8-10 kg (b) 12-15 kg
(c) 16-18 kg (d) 22-24 kg

141. Per hectare seed yield of coriander is
(a) 10-15 q (b) 18-20 q
(c) 22-28 q (d) 30-32 q

142. Spacing of coriander is
(a) 30 x 15 cm (b) 30 x 10 cm
(c) Both (d) None of these

143. Sindhu, Swathi, Sadhna, Rajendra Swathi, Pant Haritma and Azad Dhania-1 are the improved varieties of
(a) Coriander (b) Cumin
(c) Methi (d) Sauf

144. Sowing time of coriander is *(Ph.D. VGS, 2005)*
(a) August-September (b) October-November
(c) December-January (d) February-March

145. The protein content in coriander leaves is
(a) 2.1% (b) 3.3%
(c) 4.2% (d) 5.3%

146. Essential oil content in coriander is
(a) 0.1-2.7% (b) 3.2-3.8%
(c) 4.3-4.9% (d) 5.6%

147. Schizocarpic type fruit is found in
   (a) Cumin (b) Coriander
   (c) Both (d) None of these

148. Which part of coriander is used ?
   (a) Leaves (b) Seeds
   (c) Both (d) None of these

149. The chemical found in coriander is
   (a) Linalool (b) Borneol
   (c) Both (d) None of these

150. Botanical name of cumin is
   (a) *Coriandrum sativum* (b) *Cuminum cyminum*
   (c) *Foeniculun vulgare* (d) *Garania indica*

151. Cumin belongs to the family
   (a) *Solanaceae* (b) *Leguminosae*
   (c) *Apiaceae* (d) *Myrfaceae*

152. Somatic chromosome number in cumin is *(Ph.D. VGS, 2005)*
   (a) 12 (b) 14
   (c) 16 (d) 24

153. Origin of cumin is
   (a) Mediterranean region (b) China
   (c) USA (d) India

154. Sowing time of cumin is *(Ph.D. VGS, 2006)*
   (a) October (b) November
   (c) December (d) March

155. Per hectare seed rate of cumin is
   (a) 12-15 kg (b) 16-18 kg
   (c) 20-25 kg (d) 30-35 kg

156. Spacing of cumin is
   (a) 30 x 15 cm (b) 30 x 20 cm
   (c) 30 x 30 cm (d) 30 x 40 cm

157. Per hectare yield of cumin is
   (a) 5-6 q (b) 7-8 q
   (c) 9-10 q (d) 12-14 q

158. R2-19, GC-1, GJ-1, 3, 4, R2-209 and Vijapur-5 are the improved varieties of
   (a) Coriander (b) Methi
   (c) Cumin (d) Sauf

159. Cumin fruits are rich source of
   (a) Thymol (b) Limone
   (c) Eugenol (d) Anithol

160. The country which is major exporter of cumin is
    (a) Iraq (b) Iran
    (c) USA (d) China

161. Which part of cumin is used ? *(M.Sc. Ag., 2006)*
    (a) Flower (b) leaf
    (c) Fruit (d) Stem

162. The chemical found in cumin is
    (a) Cuminol (b) Eugenol
    (c) Limone (d) Anithol

163. Essential oil content in cumin is
    (a) 1-1.5% (b) 2.5-3.5%
    (c) 4.5-5.5% (d) 6.5-7.5%

164. Flower colour of cumin is
    (a) White (b) Pink
    (c) Both (d) None of these

165. Botanical name of fennel is *(Ph.D. VGS, 2006)*
    (a) *Foeniculum vulgare* (b) *Cuminum cyminum*
    (c) *Coria ndum sativum* (d) *Allium sativum*

166. Souf belongs to the family
    (a) *Apiaceae* (b) *Solanaceae*
    (c) *Malvaceae* (d) *Myrtaceae*

167. Somatic chromosome number in sauf is
    (a) 22 (b) 24
    (c) 12 (d) 32

168. Origin of fennel is
    (a) India (b) China
    (c) Mediterranean region (d) USA

169. Best sowing time of fennel is
    (a) July-August (b) September-October
    (c) November-December (d) February-March

170. Per hectare seed rate of fennel is
    (a) 6-8 kg (b) 15-20 kg
    (c) 10-12 kg (d) 25-30 kg

171. Per hectare yield of fennel is
    (a) 9-10 q (b) 11-12 q
    (c) 13-14 q (d) 15-60 q

172. RF-101, RF-125, RF-35 and Azad Sauf-1 are the improved varieties of
    (a) Cumin (b) Fennel
    (c) Clove (d) Methi

173. Which part of fennel is used ?
(a) Seed (b) Fruit
(c) Leaf (d) Stem
174. Spacing of fennel is
(a) 45 x 30 cm (b) 45 x 45 cm
(c) 45 x 60 cm (d) 60 x 45 cm
175. The chemical found in fennel is
(a) Linalool (b) Cuminol
(c) Limone (d) Eugenol
176. Essential oil content in sauf is
(a) 1-3% (b) 4-5%
(c) 6-7% (d) 8-10%
177. The chemical is responsible for sweetness in fennel is
(a) Limone (b) Anithol
(c) Eugenol (d) Cuminol
178. Flower colour of fennel is
(a) Yellow (b) White
(c) Both (d) None of these
179. Botanical name of aniseed is
(a) *Pempinella anisum* (b) *Foeniculum vulgare*
(c) *Pigmenta dioca* (d) *Vanilla planifalia*
180. Aniseed belongs to the family
(a) Solanaceae (b) Myrtaceae
(c) Apiaceae (d) Leguminosae
181. Somatic chromosome number in aniseed is
(a) 12 (b) 22
(c) 32 (d) 34
182. Origin of aniseed is
(a) China (b) India
(c) Mediterranean region (d) Europe
183. Which part of aniseed is used ?
(a) Leaves (b) Fruits
(c) Stem (d) Flower
184. The chemical found in aniseed is
(a) Anithol (b) Animone
(c) Eugenol (d) Phenol
185. Botanical name of Allspice is *(M.Sc. Ag., 2006)*
(a) *Murry Oenigi* (b) *Rimenta dioca*
(c) *Garcinia indica* (d) *Vanilla penifalia*
186. Allspice belongs to the family
(a) *Myrtaceae* (b) *Salanaceae*
(c) *Apiaceae* (d) *Leguminosae*

187. Origin of allspice is
(a) India (b) China
(c) West Indies (d) Australia

188. Jamaica is the major producer of
(a) Allspice (b) Sauf
(c) Cumin (d) Curry leaf

189. Allspice is a
(a) Monoecious plant (b) Dioecious plant
(c) Hermaphrodite plant (d) None of these

190. Fruits of allspice is called as
(a) Berry (b) Siliqua
(c) Capsule (d) Lomentum

191. Which part of allspice is used ?
(a) Leaf (b) Fruits
(c) Stem (d) flower

192. The chemical present in allspice is
(a) Phenol (b) Eugenol
(c) Anithol (d) Amimone

193. Botanical name of curry leaf is
(a) *Pimenta diola* (b) *Murrya koenigi*
(c) *Garcinia indica* (d) *Vanilla planifolia*

194. Curry leaf belongs to the family
(a) *Rutaceae* (b) *Myrtaceae*
(c) *Apiaceae* (d) *Leguminance*

195. Somatic chromosome number of curry leaf is
(a) 12 (b) 16
(c) 18 (d) 20

196. Crystaline glucoside 'Koenigine' is obtained from leaves of
(a) Allspice (b) Kokum
(c) Curry leaf (d) Vanilla

197. 'Muragin' is obtained from flower of
(a) Curry leaf (b) Sauf
(c) Aniseed (d) Allspice

198. Curry leaf is
(a) Cross pollinated (b) Self pollinated
(c) Often cross pollinated (d) None of these

199. Suwasmios a variety of
(a) Curry leaf (b) Sauf
(c) Vanilla (d) Kokum

200. Curry leaf is propagated through
(a) Leaf cuttings (b) Stem cuttings
(c) Seeds (d) None of these

201. Which part of curry leaf is used ?
(a) Leaves (b) Seeds
(c) Fruits (d) Flowers

202. The Chemical content in curry leaf is
(a) Phenol (b) Ferumine
(c) Koenigine (d) Vanillin

203. Botanical name of Kokum is
(a) *Garcinia indica* (b) *Pigmenta dioca*
(c) *Murraya koinigi* (d) *Vanilla planifalia*

204. Kokum belongs to the family
(a) *Rutaceae* (b) *Myrtaceae*
(c) *Guttiderae* (d) *Fabaceae*

205. Somatic chromosome number of kokum is
(a) 24 (b) 28
(c) 32 (d) 34

206. Origin of kokoum is
(a) China (b) USA
(c) India (d) Sri Lanka

207. Kokum is a
(a) Dioecious plant (b) Monoecious plant
(c) Gynodioecious plant (d) None of these

208. Which part of kokum is used ?
(a) Seed (b) Fruits
(c) Leaves (d) Flowers

209. The chemical found in kokum is
(a) β-hydroxy acetic acid (b) Koenigine
(c) Phenol (d) Vanilla

210. Botanical name of vanilla is
(a) *Gracinia indica* (b) *Vanilla panifalia*
(c) *Munya koenigi* (d) *Crous sativus*

211. Vanilla belongs to the family
(a) *Orchideaceae* (b) *Guttiferae*
(c) *Rutteaceae* (d) *Myrtaceae*

212. Origin of vanilla is
(a) India (b) China
(c) America (d) Europe

213. Vanilla is propagated through
(a) Stem cuttings (b) Shoot cuttings
(c) Root cuttings (d) Seeds

214. Which part of vanilla is used ?
(a) Leaves (b) Fruits
(c) Roots (d) Stem

215. The chemical present in vanilla is
   (a) Vanillin (b) Cicrocrocin
   (c) Ferumine (d) Phenol
216. Botanical name of saffron is
   (a) *Ferulla foetida* (b) *Crous sativus*
   (c) *Anethum graveolens* (d) *Nigelha sativa*
217. Saffron belongs to the family
   (a) *Iridaceae* (b) *Apiaceae*
   (c) *Umbelliferae* (d) *Leguminosae*
218. Origin of saffron is
   (a) USA (b) China
   (c) Sri Lanka (d) India
219. Leading state for saffron cultivation is
   (a) Kashmir (b) Himachal Pradesh
   (c) Andhra Pradesh (d) Kerala
220. Saffron is propagated through
   (a) Seeds (b) Bulbs
   (c) Corms (d) Rhizomes
221. Which part of saffron is used ?
   (a) Dried stigma (b) Anthers
   (c) Flowers (d) Seeds
222. Which chemical is found in saffron ?
   (a) Vanillin (b) Phenol
   (c) Circrocrocin (d) Ferumine
223. The amount of essential oil content in saffron is
   (a) 0.5-1% (b) 2-2.5%
   (c) 3.5-4% (d) 5-6%
224. Botanical name of Asafaetida is
   (a) *Crous sativus* (b) *Ferulla foetida*
   (c) *Anethum graveolens* (d) *Nigella sativa*
225. Heeng belongs to the family
   (a) Apiaceae (b) *Poaceae*
   (c) *Iridaceae* (d) *Euphoribiaceae*
226. Himachakta is prepared from
   (a) Saffron (b) Ajowain
   (c) Heeng (d) Dill
227. Which part of asafoetida is used ?
   (a) Oleoresingum (b) Fruit
   (c) Flower (d) Root
228. The chemical present in Heeng is
   (a) Circrocrocin (b) Carvone
   (c) Ferumine (d) Phenol

229. The amount of essential oil content in Heeng is
(a) 10-17% (b) 19-22%
(c) 23-28% (d) 29-35%

230. Botanical name of ajowain is
(a) *Ferulla foetida* (b) *Trachyspermum ammi*
(c) *Anethum graveolens* (d) *Nigella sativa*

231. Ajowain belongs to the family
(a) *Apiaceae* (b) *Linaceae*
(c) *Iridaceae* (d) *Solanaceae*

232. Origin of ajowain is *(M.Sc. Ag., 2006)*
(a) South-West Asia (b) South East Asia
(c) Europe (d) USA

233. Which part of ajowain is used ?
(a) Fruits (b) Flowers
(c) Leaves (d) Roots

234. Per hectare seed rate of ajowain is
(a) 1-2 kg (b) 3-4 kg
(c) 5-6 kg (d) 7-8 kg

235. Sowing space of ajowain is
(a) 30 x 15 cm (b) 30 x 30 cm
(c) 30 x 45 cm (d) 45 x 20 cm

236. Sowing time of ajowain is
(a) August-September (b) October-November
(c) December-January (d) February-March

237. Botanical name of dill is
(a) *Anethum graveolens* (b) *Tranchyspermum ammi*
(c) *Nigella sativa* (d) *Crocus sativus*

238. Dill belongs to the family *(M.Sc. Ag., 2006)*
(a) *Leguminosae* (b) *Euphorbiaceae*
(c) *Apiaceae* (d) *Solanaceae*

239. Which part of soya is used ?
(a) Seed (b) Root
(c) Leaves (d) Flower

240. The chemical found in soya is
(a) Phenol (b) Ferumine
(c) Carvone (d) Anithol

241. Essential oil content in soya is *(M.Sc. Ag., 2006)*
(a) 1.5-4% (b) 5-6%
(c) 7-9% (d) 11-12%

242. Which spice seeds are used for making gripe water ?
(a) Sauf (b) Soya
(c) Ajowain (d) Saffron

243. Sowing time of soya is
(a) October (b) November
(c) December (d) January

244. Flower colour of soya is
(a) White (b) Green
(c) Yellow (d) Red

245. Per hectare seed rate of soya is
(a) 3-5 kg (b) 7-9 kg
(c) 10-12 kg (d) 14-16 kg

246. Botanical name of nigella is
(a) *Anethum graveolous* (b) *Nigella sativa*
(c) *Crocus sativus* (d) *Ferulla foetida*

247. Nigella belongs to the family *(M.Sc. Ag., 2006)*
(a) *Apiceae* (b) *Ramiculaceae*
(c) *Solanaceae* (d) *Fabaceae*

248. Origin of Nigella is
(a) South –East Asia and America
(b) China
(c) India (d) Sri Lanka

249. Per hectare seed rate of Nigella is
(a) 5 kg (b) 10 kg
(c) 15 kg (d) 20 kg

250. Sowing time of Nigella is *(Ph.D. VGS, 2006)*
(a) October-November (b) December-January
(c) February-March (d) May-June

251. Azad Kalonji-1 is the improved variety of
(a) Dill (b) Nigella
(c) Ajowain (d) Saffron

252. Per hectare yield of Nigella is
(a) 10-15 q (b) 20-25 q
(c) 30-35 q (d) 40-45 q

253. How many spices are grown in the world ?
(a) 107 (b) 95
(c) 50 (d) 102

254. How many spices are grown in India ?
(a) 18 (b) 35
(c) 38 (d) 52

255. How many seed spices are grown in India ?
(a) 5 (b) 3
(c) 7 (d) 9

256. How many spices are there where floral parts are used ?
   (a) 2 (b) 5
   (c) 8 (d) 7
257. The rhizome spices are
   (a) 2 (b) 4
   (c) 3 (d) 5
258. The berry spices are
   (a) 2 (b) 3
   (c) 4 (d) 5
259. The fruit spices are
   (a) 2 (b) 3
   (c) 4 (d) 5

ANSWERS

1. (b)	2. (b)	3. (a)	4. (a)	5. (b)
6. (a)	7. (b)	8. (a)	9. (c)	10. (a)
11. (a)	12. (b)	13. (a)	14. (b)	15. (a)
16. (b)	17. (a)	18. (b)	19. (a)	20. (a)
21. (a)	22. (b)	23. (b)	24. (b)	25. (a)
26. (a)	27. (b)	28. (b)	29. (a)	30. (b)
31. (a)	32. (b)	33. (c)	34. (a)	35. (b)
36. (a)	37. (a)	38. (b)	39. (b)	40. (a)
41. (b)	42. (b)	43. (c)	44. (a)	45. (b)
46. (a)	47. (c)	48. (a)	49. (a)	50. (b)
51. (a)	52. (b)	53. (a)	54. (b)	55. (a)
56. (a)	57. (b)	58. (b)	59. (c)	60. (a)
61. (b)	62. (b)	63. (b)	64. (a)	65. (a)
66. (b)	67. (a)	68. (b)	69. (a)	70. (a)
71. (a)	72. (b)	73. (a)	74. (b)	75. (a)
76. (b)	77. (c)	78. (a)	79. (c)	80. (b)
81. (b)	82. (b)	83. (a)	84. (b)	85. (a)
86. (b)	87. (a)	88. (b)	89. (a)	90. (b)
91. (a)	92. (c)	93. (b)	94. (a)	95. (b)
96. (c)	97. (a)	98. (b)	99. (b)	100. (a)
101. (c)	102. (c)	103. (b)	104. (a)	105. (b)
106. (c)	107. (c)	108. (a)	109. (b)	110. (a)
111. (c)	112. (c)	113. (b)	114. (b)	115. (a)
116. (b)	117. (c)	118. (b)	119. (a)	120. (b)
121. (b)	122. (a)	123. (b)	124. (b)	125. (a)
126. (b)	127. (b)	128. (a)	129. (b)	130. (a)

131. (b)	132. (c)	133. (c)	134. (a)	135. (c)
136. (a)	137. (b)	138. (c)	139. (b)	140. (b)
141. (b)	142. (c)	143. (a)	144. (b)	145. (b)
146. (a)	147. (b)	148. (c)	149. (c)	150. (b)
151. (c)	152. (b)	153. (a)	154. (b)	155. (a)
156. (a)	157. (b)	158. (c)	159. (a)	160. (b)
161. (c)	162. (a)	163. (b)	164. (c)	165. (a)
166. (a)	167. (a)	168. (c)	169. (b)	170. (c)
171. (a)	172. (b)	173. (b)	174. (a)	175. (c)
176. (a)	177. (b)	178. (a)	179. (a)	180. (c)
181. (b)	182. (c)	183. (b)	184. (a)	185. (b)
186. (a)	187. (c)	188. (a)	189. (b)	190. (a)
191. (b)	192. (a)	193. (b)	194. (a)	195. (c)
196. (c)	197. (a)	198. (b)	199. (a)	200. (c)
201. (a)	202. (c)	203. (a)	204. (c)	205. (a)
206. (c)	207. (a)	208. (b)	209. (a)	210. (b)
211. (a)	212. (c)	213. (a)	214. (b)	215. (a)
216. (b)	217. (a)	218. (d)	219. (a)	220. (c)
221. (a)	222. (c)	223. (a)	224. (b)	225. (a)
226. (c)	227. (a)	228. (c)	229. (a)	230. (b)
231. (a)	232. (b)	233. (a)	234. (b)	235. (a)
236. (b)	237. (a)	238. (c)	239. (a)	240. (c)
241. (a)	242. (b)	243. (a)	244. (c)	245. (a)
246. (b)	247. (b)	248. (a)	249. (b)	250. (a)
251. (b)	252. (a)	253. (a)	254. (d)	255. (c)
256. (a)	257. (a)	258. (a)	259. (a)	

CHAPTER 12

OKRA

- Okra belongs to the family *Malvaceae* and India is the largest producer in the world.
- Okra is more remunerative than the leafy vegetables.
- The root and stem of okra are useful for cleaning cane juice in the preparation of jaggery.
- The species *Abelmoschus manihot* is also used as leafy vegetable.
- Bhindi flowers are self-fertile, but usually upto 10% cross-pollination occurs by insects, thus it is classified as an often-cross-pollinated crop.
- Dr. Harbhajan Singh initiated systemic research work on improvement of *bhindi* and he developed Pusa Sawani the first okra variety which became most popular.
- Pusa A-4 has excellent shelf-life. It is resistant to YVMV and tolerant to jassids and shoot and fruit borers.
- Arka Anamika variety has been developed as a hybrid derivative from inter-specific cross between *A. esculentus* and A. *tetraphyllus.*
- Panjab Padmini variety has been developed as a hybrid derivative from inter-specific cross between *A. esculentus* and *A. manihot ssp manihot.*
- Prabhani Kranti, a YVMV resistant variety, was developed from inter-specific cross.
- The variety, Perkin's Long Green was released in 1983-84 as Harbhajan Bhindi from College of Agriculture, Solan to commemorate the memory of Dr. Harbhajan Singh.
- At temperature above 42°C, flower buds in most of the cultivars of okra may desiccate and drop causing yield losses.
- *Bhindi* does not require training or pruning. Varieties developed so far are upright growing and hence staking is also not required.
- YVMV (Yellow Vein Mosaic Virus) is the most serious disease of okra. Its vector is white fly (*Bemisia tabaci*).
- Jassid is the most serious pest of okra.

- Post-harvest treatment with cycocel (100 ppm) enhances shelf-life of fruits.
- For processing industry and fresh fruit export 6-8 cm long fruits are desirable.

The objective type questions on okra are as follows :

1. Okra is more remunerative than the vegetables
   (a) Cucurbitaceous (b) Leafy
   (c) Solanaceous (d) None of these
2. *Abelmoschus esculentus* (L.)Moench. is the botanical name of
   (a) Bhindi (b) Palak
   (c) Methi (d) None of these
3. The root and stem of okra are used for
   (a) Cleaning cane juice (b) Preparation of jaggery
   (c) Both (d) None of these
4. Okra species *Abelmoschus manihot* is used as
   (a) Fruit vegetable (b) Leafy vegetable
   (c) Both (d) None of these
5. For fresh okra export, fruits should be
   (a) Green (b) Tender
   (c) 6-9 cm long (d) All the above
6. Suitable varieties of okra for export are
   (a) Pusa Sawani (b) Parbhani Kranti
   (c) Varsha Uphar, Pusa-4 (d) All the above
7. Bhindi flowers are self-fertile, but usually cross pollination occurs by insects upto
   (a) 4% (b) 5%
   (c) 8% (d) 10%
8. Okra is a
   (a) Self pollinated plant (b) Cross pollinated plant
   (c) Often cross pollinated plant
   (d) None of these
9. The systemic research work on improvement of bhindi was started by
   (a) Puskar Nath (b) B. Choudhary
   (c) Harbhajan Singh (d) None of these
10. Excellent shelf-life variety of okra is
    (a) Pusa Sawani (b) Prabhani Kranti
    (c) Pusa A-4 (d) None of these
11. Okra variety resistant to YVMV and tolerant to Jassids and Shoot and fruit borers is
    (a) Pusa Sawani (b) Azad B-1
    (c) Pusa A-4 (d) None of these

12. MDU-1 is an induced mutant from Pusa Sawani by using
    (a) Alfa rays (b) β-rays
    (c) Gamma rays (d) None of these
13. Arka Anamika variety has been developed as a hybrid derivative selection from inter-specific cross between *A. esculentus* and ______.
    (a) *A. manihot* (b) *A. tetra phyllus*
    (c) *A. Penta phyllus* (d) None of these
14. Punjab Padmini variety has been developed as a hybrid derivative selection from inter-specific cross between *A. esculentus* and ______.
    (a) *A. manihot* (b) *A. tetra phyllus*
    (c) *A. Penta phyllus* (d) None of these
15. Punjab-7 has been developed as a hybrid derivative selection of a cross between *A. esculentus* cv-Pusa Sawani and ______.
    (a) tetraphyllus (b) manihot
    (c) Both (d) None of these
16. Punjab-8 is an induced mutant derived from Pusa Sawani by treating seed with EMS with the per cent concentration of
    (a) 1 (b) 2
    (c) 3 (d) 4
17. Parbhani Kranti a YVMV resistant variety, was developed through
    (a) Intra-specific (b) Inter-specific crossing
    (c) Hybridization (d) None of these
18. The variety Perkins Long Green was released in 1983-84 as Harbhajan Bhindi from College of Agriculture, Solan to Commemorate the memory of ______.
    (a) V. Swarup (b) Y.R. Mehta
    (c) Y.S. Parmar (d) Harbhajan Singh
19. At the temperature above ______ °C, flower buds in most of the cultivars of okra may desiccate and drop causing yield losses
    (a) 30 (b) 40
    (c) 42 (d) 50
20. Salinity tolerant variety of okra is
    (a) Arka Anamika (b) Pusa Sawani
    (c) Punjab-7 (d) Azad Bhindi-2
21. About 5,000 plants per hectare are recommended for branching types and ______ plants per hectare for non-branching types
    (a) 5,500 (b) 6,000
    (c) 6,600 (d) None of these

22. Varieties like ______ and ______ give quick branching after pruning
    (a) Arka Abhay, Pusa A-4 (b) Pusa A-4, AB-1
    (c) AB-1, AB-2 (d) None of these
23. Yellow Vein Mosic Virus is most serious disease of okra and its vector is
    (a) Jassids (b) White fly
    (c) Both (d) None of these
24. The most serious pest of okra is
    (a) Tabaci (b) Hoppers
    (c) Jassids (d) None of these
25. Okra seed treatment by GA (400 ppm, IAA 20 ppm or NAA 20 ppm) enhances
    (a) Germination (b) Flowering
    (c) Fruit yield (d) None of these
26. Cycocel (1,000-1,500 ppm) reduces
    (a) Plant height (b) Fruit size
    (c) Germination (d) None of these
27. Post-harvest treatment with cycocel (100 ppm) enhances
    (a) Shelf-life of fruits (b) Dormancy
    (c) Both (d) None of these
28. The *A. esculentus* (L.) Moench is botanical name of okra in place of *H. esculentus* (L.) is due to persistant nature of
    (a) Sepals (b) Petals
    (c) Both (d) None of these
29. The mucilage in okra is due to
    (a) Carbohydrates (b) Protein
    (c) Vitamins (d) None of these
30. The YVMV disease of okra is genetically controlled with number of genes
    (a) One (b) Two
    (c) Both (d) None of these
31. Red fruit of okra is found in
    (a) AB-1 (b) Punjab Padmini
    (c) K.S.-305 (d) None of these
32. The origin of okra is
    (a) South Africa (b) India
    (c) China (d) None of these
33. The chromosome number in okra is
    (a) 2n = 16 (b) 2n = 130
    (c) 2n = 64 (d) 2n = 140

34. Fruit of okra is rendered unmarketable due to the attack of *(ARS, 2003)*
    (a) Aphids (b) Mites
    (c) Painted bug (d) None of these
35. Okra variety Punjab-8 has been evolved by *(ARS, 2003)*
    (a) Inter-specific hybridization
    (b) Irradiation induced mutation
    (c) Chemical induced mutation
    (d) None of the above
36. The anthesis in okra takes place from *(ARS, 2003)*
    (a) 8 to 10 a.m. (b) 5 to 7 a.m.
    (c) 1 to 2 p.m. (d) 3 to 5 p.m.
37. The minimum temperature required for germination of seed of okra is *(ARS, 2003)*
    (a) 10°C (b) 20°C
    (c) 30°C (d) None of these
38. The most serious disease of okra is *(ARS, 2003)*
    (a) Yellow vein mosaic (b) Powdery mildew
    (c) Root rot (d) None of the above
39. First picking in okra can be done after *(ARS, 2003)*
    (a) 30-35 days (b) 45-50 days
    (c) 60-65 days (d) None of these
40. In *kharif* season the okra crop is sown at a spacing of *(Ph.D. VGS, 2006)*
    (a) 75 x 45 cm (b) 60 x 45 cm
    (c) 45 x 15 cm (d) None of these
41. Which of the following is a variety of okra? *(Ph.D. VGS, 2006)*
    (a) Pusa Makhamali (b) Sugar Baby
    (c) Arka Punit (d) Arka Manik
42. Anthesis in okra takes place from *(Ph.D. VGS, 2006)*
    (a) 5 to 7 a.m. (b) 8 to 10 a.m.
    (c) 3 to 5 p.m. (d) None of these
43. "Parbhani Kranti" is a popular variety of *(UP PCS (Pre), 2003)*
    (a) Brinjal (b) Okra
    (c) Potato (d) Chilli
44. Which of the following varieties of okra is suitable for hills? *(ARS, 1992)*
    (a) Pusa Sawani (b) Pusa Makhamali
    (c) Perkin's Long Green (d) None of these

45. In central and southern India okra can be grown in *(ARS, 1992)*
    (a) Summer (b) Winter
    (c) Spring (d) Throughout year
46. Yellow leaf mosaic of bhindi crop spreads by *(UP PCS (Pre), 2005)*
    (a) Jassids (b) Borers
    (c) Jassids and Borers both (d) None of these
47. Okra variety developed through mutation is *(ARS, 1999)*
    (a) MDU-1 (b) Azad Bhindi-1
    (c) PKM-1 (d) None of these
48. The place of origin of okra is (ARS, 1999)
    (a) South America (b) Tropical Africa
    (c) Peru (d) None of these
49. Which of the following variety belongs to okra ? *(ARS, 1999)*
    (a) Sugar Baby (b) Azad B-1
    (c) Arka Punit (d) None of these
50. *Varsha* Upahar is the variety of *(Ph.D. VGS, 2004)*
    (a) Chilli (b) Okra
    (c) Cabbage (d) None of these
51. The variety of bhindi is *(Ph.D. VGS, 2005)*
    (a) Pusa Ruby (b) Arka Manik
    (c) Arka Komal (d) Arka Anamika
52. *Versa* Bahar is an improved variety of
    (a) Okra (b) Chilli
    (c) Cowpea (d) Guar

ANSWERS

1. (b)	2. (a)	3. (a)	4. (b)	5. (d)
6. (d)	7. (d)	8. (c)	9. (c)	10. (c)
11. (c)	12. (c)	13. (b)	14. (a)	15. (b)
16. (a)	17. (b)	18. (d)	19. (c)	20. (b)
21. (c)	22. (a)	23. (b)	24. (c)	25. (a)
26. (a)	27. (a)	28. (a)	29. (a)	30. (a)
31. (c)	32. (a)	33. (b)	34. (c)	35. (c)
36. (a)	37. (b)	38. (a)	39. (b)	40. (c)
41. (a)	42. (b)	43. (b)	44. (c)	45. (d)
46. (c)	47. (a)	48. (b)	49. (b)	50. (b)
51. (d)	52. (a)			

CHAPTER 13

AGRICULTURAL STATISTICS

Definition of statistics

The science of statistics is the study of principles and methods used in the collection, analysis and interpretation of numerical data in any sphere of enquiry. Statistics are the numerical statements of facts capable of being analyzed and interpreted.

There are the three aims of statistics *viz.,* study the population, study the variation and its causes and to study the methods of reduction of data.

The important limitations of statistics are as given below:

(1) It does not study qualitative phenomena.

(2) Statistical studies are true only on an average.

(3) It does not study individuals.

(4) It does not reveal the entire story and

(5) It is liable to be misused.

13.1 Measures of central tendency

Central tendency may be defined as a value of the variate which is thoroughly representative of the series or the distribution as a whole. The measure of this character is known as averages.

13.1.1 Average

The following are the different types of averages which are in common use:

(1) Arithmetic Mean

(2) Median

(3) Mode

(4) Geometric Mean

(5) Harmonic Mean

(6) Weighted Mean

13.2 Standard Deviation

The calculation standard deviation is based on the deviations from the arithmetic mean as

$$S.D. = \sqrt{\frac{\Sigma(x-\mu)^2}{N}}$$

Where,

x = An observation or variate value

μ = Arithmetic mean of the population

N = Number of given observations

13.3 Variance

Variance is the square of the standard deviation variance $= (S.D.)^2$

13.4 Coefficient of Variance

This is also a relative measure of dispersion. It is important on account of the widely used measure of central tendency and dispersion i.e., Arithmetic Mean and SD.

It is given by –

$$C.V. = \frac{S.D.}{A.M.} \times 100$$

13.5 Correlation

If the change in one variable is accompanied by a change in the other variable in such a way that an increase in one variable results in an increase or decrease in the other and also greater change in one variable results in a corresponding greater change in other, then the two variables are said to be correlated.

13.6 Experimental Design

(1) Treatments: The objects of comparison, which an experimenter has to try in the field for assessing their values are known as treatments e.g., the varieties, manures, cultivation practices, etc.

(2) Experimental units: Experimental material is divided into a number of ultimate units to which the treatments are applied and are known as experimental units.

(3) Experimental material: This is the name given to the material on which the experiment is performed e.g., a field, soil, patients in a hospital, etc.

13.6.1 Basic Principles of Field Experimentation

There are following three basic principles of field experimentation:

(1) Replication: The allocation of the treatments under investigation is known as replication.

(2) Randomization: The allocation of the treatments to the different experimental units by a random process is known as randomization of treatments.

(3) Local control: The principles of making use of greater homogeneity in groups of experiment (e.g., blocks of a number of plots homogeneous within themselves) for reducing the experimental error as explained in the foregoing paragraph is known as local control. This is done by dividing the experimental material (e.g., field) into homogeneous groups of experimental units (e.g., blocks).

13.6.2 Conditions for application of designs

Completely Randomized Design (C.R.D.) is applied when the experimental material is limited and homogeneous, such as the soil in the pot experiments. The advantages of this design are :

(1) The method of analysis remains simple when the results from some units are rejected.

(2) This design is specially useful in small experiments where, the supply of experimental material is scarce and homogeneous.

(3) In this design any number of treatments and replicates may be used.

(4) The statistical analysis of the data is very easy.

(5) The relative loss of information due to missing data is smaller than that with any other design.

Randomized Block Design (R.B.D.) is appropriate when the fertility gradient of the field is in one direction. It may be adopted upto 20 treatments without an appropriate loss of efficiency.

(1) In R.B.D., the number of blocks must be equal to the number of replications fixed for each treatment.

(2) The number of plots in each block should be equal to the number of treatments.

(3) Randomized of treatments in each block should be a fresh.

(4) The experimental error within each block is to be kept as small as practically possible and the variation from block to block as great as possible. In this way all the treatments which are assigned to one block and therefore the same type of environmental effects and therefore comparable.

Latin Square Design (L.S.D.) can replace the R.B.D., when the fertility gradient is in two directions, instead of one. It may be adopted for number of treatments ranging from 5 to 8 or from 5 to 12 at the most. This should not be used when the treatments are less than 5.

The important points about L.S.D. are :

(1) No. of replications = No. of treatments.

(2) No. of rows = No. of columns = No. of treatments.

(3) Randomization of treatments is done in such a way that each treatment occurs once and only once in each row and each column.

13.7 Test of Significance

The statistical procedure for deciding whether the difference under study is significant or non-significant, is called the test of significance.

(a) Null hypothesis: The hypothesis which is a test for possible rejection under assumption that it is true.

(b) Alternate hypothesis: Any hypothesis which is complementary to the null hypothesis is called alternate hypothesis. This is denoted by H_1.

13.8 Two Types of Error

1. Type I error: Type I error is committed when we reject the hypothesis when it is true. The probability of committing a type I error is denoted by α.
2. Type II error: Type II error is committed by not rejecting the null hypothesis when it is false. The probability of committing a type II error is denoted by β.

13.9 Level of Significance

The maximum probability of rejecting the hypothesis when it is true or in other words, the maximum probability of type I error is known as level of significance.

13.10 x^2 (Chi-square)-test

The value of the quantity of x^2 is calculated by the help of the following expression:

$$x^2 = \Sigma\frac{(O - E)^2}{E}$$

Where, O = Observed frequency in a cell

E = Expected frequency in a cell

Σ = Summation

13.11 General

- Standard error and standard deviation are synonymous.
- The coefficient of variation is often between 5 and 15 per cent.
- Randomization does not guarantee to balance out the natural difference between the members of a pair exactly.

- Randomization gives each treatment an equal chance of being allotted to any subject giving unusually good or poor response exactly as assumed in the theory of probability.
- Range-test like t-test assumes a normal distribution and has become popular, particularly in industrial work.
- The correlation coefficient (r) is a measure of the degree of closeness between two variables.
- The study of probability began around three hundred years ago.
- The variance ratio (F) was first tabulated by Fisher in whose honour the criterion was named F by Snedecor.

The essential feature of the split-plot experiment is that the sub-plot treatments are not randomized over the whole large block, but only over the main plots.

The objective type questions on agricultural statistics are given below :

1. The standard deviation of the following values 15, 15, 15, 15, 15 is
   (a) 15 (b) 5
   (c) 0 (d) None
2. An Experimental design is
   (a) A plan of experiment (b) A map
   (c) Both (d) None of these
3. An Experiment Unit in a research work is
   (a) A group of insects (b) A field plot
   (c) An animal (d) All the above
4. Randomization is a process in which the treatments are allocated to the experimental units
   (a) With equal probability (b) In a sequence
   (c) Both (d) None of these
5. Replication is an experiment provides
   (a) The estimate of experimental error
   (b) Check the variation
   (c) Both (d) None of these
6. Number of replications in an experiment is based on
   (a) The experimental material available
   (b) The precision required
   (c) Heterogeneity of experimental material
   (c) All the above
7. With the help of contrast one can estimate the
   (a) Cubic effect (b) Quadratic effect
   (c) Linear effect (d) All the above

8. The idea of preliminary test of significance was propounded by
   (a) T.A. Bancrofit (b) A.E. Paull
   (c) F. Yates (d) None of these
9. A latin square design possesses
   (a) One way classification (b) Two way classification
   (e) Three way classification (d) None of these
10. In an experiment with 4 varieties and 3 fertilizers, the degrees of freedom for the interaction V x F are
    (a) 4 (b) 6
    (c) 12 (d) None of these
11. The average height of plant is 50 cm and S.D. is 10 cm the coefficient of variance will be
    (a) 80% (b) 15%
    (c) 30% (d) 20%
12. Which transformation is most appropriate for percentage ?
    (a) Square root (b) Arc sine
    (c) Logarithmic (d) All of these
13. If any test of significance, type I error is committed when
    (a) A true hypothesis is accepted
    (b) A true hypothesis is rejected
    (c) A false hypothesis is accepted
    (d) A false hypothesis is rejected
14. The coefficient of correlation 'r' of the characters 'X' and 'Y' is
    (a) $r = \sqrt{byx/bxy}$ (b) $r = \sqrt{byx\text{-}bxy}$
    (c) $r = \sqrt{byx \,.\, bxy}$ (d) None of these
15. The mean and median of a series are 79.77 and 79.06, respectively. The mode of the series is
    (a) 77.64 (b) 75.80
    (c) 76.64 (d) 72.80
16. If two observations are +5 and –5, their geometric mean is
    (a) 5 (b) –25
    (c) –4 (d) None of these
17. If the regression coefficients are b1 and b2 then the correlation coefficient 'r' will be
    (a) b1 / b2 (b) b2 / b1
    (c) b1 / b2 (d) $\pm\sqrt{b1 \times b2}$
18. If P1 and P2 are the probabilities of a happening of two mutually exclusive events A and B, then the probability of happening of both A and B will be
    (a) 1 (b) 2
    (c) 0 (d) None of these

19. The two frequencies a and a^2 are expected to be in the ratio of 3:2, to test this hypothesis, the value of x^2 will be calculated by
    (a) $(3a1 - 2a2)^2 / 6(a1 + a2)$ (b) $(2a1\text{-}3a2)^2/5(a1+a2)$
    (c) $(2a1 - 3a2)^2 / 6(a1 + a2)$ (d) None of these
20. If byx =0.16 and bxy=0.09 then correlation coefficient will be
    (a) 1.20 (b) 0.12
    (c) –0.12 (d) None of these
21. Which one is the measurement of the mean
    (a) MD (b) HM
    (c) GM (d) None of these
22. What is the SD (Sum of deviation) of a set of 50 numbers from their arithmetic mean
    (a) 100 (b) 50
    (c) 25 (d) 0
23. In confounding design
    (a) Main effects are confounded
    (b) Higher order contrast is confounded
    (c) Lower order contrast is confounded
    (d) None of these
24. The method of confounding is a device to reduce the size of
    (a) Experiments (b) Replications
    (c) Blocks (d) All the above
25. If the same factorial effect is confounded in all the replications, it is known as
    (a) Complete confounding (b) Partial confounding
    (c) Both (d) None of these
26. In 2^3 factorial, all effects and their sum of squares can be obtained directly by
    (a) Yates method (b) Modulo technique
    (c) Both (d) None of these
27. If there is a factor at four levels in a factorial experiment, one can find its
    (a) Linear effect (b) Cubic effect
    (c) Quadratic effect (d) All the above
28. The main and interaction effects in a 2^3 factorial can easily be estimated with the help of
    (a) Simple effect (b) Contrasts
    (c) Both (d) None of these

29. The effect, which is confounded in all the blocks in an experimental design
    (a) Is estimated more precisely
    (b) Is estimated less precisely
    (c) Can not be estimated (d) None of these
30. If different effects are confounded in different blocks, it is said to be
    (a) Complete confounding (b) Partial confounding
    (c) Balance confounding (d) None of these
31. If all effects of the same order are confounded with incomplete blocks differences, it is said to be
    (a) Complete confounding (b) Partial confounding
    (c) Balance confounding (d) None of these
32. If the interactions AB and BC are confounded with in complete block in 2^3 factorial experiment, then automatically confounded effect is
    (a) A (b) B
    (c) ABC (d) AC
33. The factors having the levels equal to the quotient which is obtained on dividing the total number of treatment combinations of an asymmetrical factorial by the incomplete block size are called
    (a) Real factors (b) Pseudo factors
    (c) Both (d) None of these
34. The factors which are used to convert the asymmetrical factorial into symmetrical factorial for the purpose of confounding are termed as
    (a) Real factors (b) Pseudo factors
    (c) Both (d) None of these
35. The factors, not having the level (S) equal to the quotient which is obtained on dividing the total number of treatment combinations by the incomplete block size, are called
    (a) The factors of asymmetry (b) Real factors
    (c) Pseudo factors (d) None of these
36. Confounding in asymmetrical factorial is carried out with the help of
    (a) Modulo technique (b) Contrasts
    (c) Both (d) None of these
37. When the treatments require different size plot for experimentation, frequently adopted design is
    (a) Split plot design (b) Confounding design
    (c) Latin squre design (d) None of these

38. In 3^3 factorial with factors A, B and C each at 3 levels, the interaction $A^2 BC^2$ is same as the interaction
    (a) ABC (b) AB^2C
    (c) $A B^2 C^2$ (d) None of these
39. An experiment in which all factors have different levels, is known as
    (a) Typical factorial (b) Asymmetrical factorial
    (c) Pseudo factorial (d) None of these
40. The experiments with various factors having unequal number of levels are called
    (a) Symmetrical factorial (b) Asymmetrical factorial
    (c) Typical factorial (d) None of these
41. The additional effect gained due to combined effect of two or more factors is known as
    (a) Main effect (b) Interaction effect
    (c) Both (d) None of these
42. Two types of effects measured in a factorial experiment are
    (a) Main and interaction effects
    (b) Simple and common effects
    (c) Both (d) None of these
43. An experiment having several factors with equal number of levels is known as
    (a) Asymmetrical experiment (b) Symmetrical experiment
    (c) Factorial experiment (d) None of these
44. The experiments in which the effects of the levels of a factor are considered at various levels of the factor are called
    (a) Symmetrical experiments (b) Factorial experiments
    (c) Rotational experiments (d) All the above
45. Error degrees of freedom for a change over design are more when the subjects are taken
    (a) Replicates (b) A group of latin square
    (c) Any order (d) None of these
46. Cross over design is also known as
    (a) Change over design (b) Reversal design
    (c) Switch back design (d) All the above
47. Cross over design is suitable for measuring
    (a) Direct treatment effect
    (b) Treatments residual effect
    (c) Both (d) None of these
48. A Latin square design controls
    (a) Multi way variation (b) Three way variation
    (c) Two way variation (d) All the above

49. The maximum possible number of orthogonal contrasts among four treatments is
(a) 1 (b) 2
(c) 3 (d) 4

50. Which of the following is not contrast among three treatments ?
(a) $T1 + 2T_2 - T_3$ (b) $T_1 - T_3$
(c) $T_2 - 2T_2 + T_3$ (d) None of these

51. Which of the following is a contrast ?
(a) $3T_1 + T_2 - 3T_2 + T_2$ (b) $T_1 + 3T_2 - 3T_3 + T_4$
(c) $-3T_1 + T_1 - T_3 + 3T_4$ (d) $T_1 + T_2 - T_3 + T_4$

52. The contrast representing the quadratic effect among four treatments is
(a) $-3T_1 - T_2 + T_3 + 3T_4$ (b) $-T_1 + 3T_2 - 3T_3 + T_4$
(c) $T_1 - T_2 - T_3 + T_4$ (d) $T_1 + T_2 + T_3 + T_4$

53. The contrast representing the line effect among four treatments is
(a) $T1 - 2T_2 + T_3$ (b) $T_1 - T_2 - T_3 + T_4$
(c) $-T_1 + 3T_2 - 3T_3 + T_4$ (d) None of these

54. The cubic effect among five treatments can be estimated by the contrast
(a) $-T_1 + 2T_2 - 2T_4 + T_5$ (b) $2T_1 - T_2 - 2T_3 - T_4 + 2T_5$
(c) $-2T_1 - T_2 + T_4 + 2T_5$ (d) None of these

55. Each contrast among K treatments has
(a) One d.f. (b) (K-1) d.f.
(c) K.d.f. (d) None of these

56. An experimental design which does not involve a randomization process is known as
(a) Non-randomized design (b) Systematic design
(c) Asymmetrical design (d) None

57. According to W.G. Cochram, error mean square in case of RBD as compared to CRD utilising same experimental material is
(a) 40% (b) 60%
(c) 80% (d) 100%

58. Who is considered as father of statistics ?
(a) RA Fisher (b) Pearson
(c) AL Bowley (d) Boddington

59. When statistics is used in plural sense it means
(a) Statistical methods (b) Collect statistics
(c) Both (d) None of these

60. Rank correlation is tested by
(a) Quade test (b) Friedmann test
(c) t-test (d) F-test

61. Which measure of central tendency is applied for calculation of regression and correlation coefficient ?
(a) Arithmetic mean (b) Mode
(c) Median (d) Geometric mean

62. The intelligency ability or efficiency can be measured by the use of
(a) Median (b) Mode
(c) Weighted HM (d) Geometric mean

63. The most frequent by occurred value of the data whose frequency is maximum is known as
(a) Arithmetic mean (b) Mode
(c) Median (d) GM

64. Which measure of central tendency is especially suited in the field of business ?
(a) Median (b) GM
(c) Harmonic mean (d) Mode

65. In a symmetrical distribution of the data which pair is correct ?
(a) Mean = Median = Mode (b) Mean > Median > Mode
(c) Mean < Median < Mode (d) Mean = Median > Mode

66. Geometric mean of a given series is always less than its
(a) Harmonic mean (b) Weighted mean
(c) Arithmetic mean (d) Weighed harmonic mean

67. If any items of the series is zero, which mean of central tendency becomes zero
(a) Harmonic mean (b) Geometric mean
(c) Arithmetic mean (d) Weighed harmonic mean

68. When we want average of rates of change or ratios or index number, which measure of central tendency is suitable ?
(a) Geometric mean (b) Mode
(c) Median (d) Arithmetic mean

69. To calculate the average speed, which item of central tendency is most suited
(a) Mode (b) Median
(c) Weighted HM (d) GM

70. Which type of average would be suitable to calculate the size of agricultural holding ?
(a) Mode (b) Median
(c) Weighted HM (d) Weighed AM

71. Which is the simplest measure of dispersion ?
(a) Mean deviation (b) Standard deviation
(c) Range (d) Variance

72. The variance is the square of
(a) Mean deviation (b) Standard deviation
(c) Range (d) Correlation

73. Mean deviation is the least when calculated about (Minimal property)
(a) Arithmetic mean (b) Median
(c) Mode (d) Geometric mean

74. If all the variate values are the same (e.g., 8, 8, 8) the standard deviation will be
(a) 4 (b) 8
(c) 16 (d) 0

75. The coefficient of variation can be calculated by using the formula
(a) Standard deviation/arithmetic mean x 100
(b) Arithmetic mean/standard deviation x 100
(c) Square root of standard deviation/arithmetic mean x 100
(d) Standard deviation/geometric mean x 100

76. Unit less measure of dispersion is
(a) Standard deviation (b) Mean deviation
(c) Range (d) CV

77. The second central moment (about the arithmetic mean) is always equal to
(a) Variance (b) Mean deviation
(c) Standard deviation (d) None of these

78. The first central moment (μ_1) is always
(a) 0 (b) 2
(c) 4 (d) 6

79. When a frequency distribution is not symmetrical about the mean, it is said to be
(a) Kurtosis (b) Moment
(c) Skewed (d) None of these

80. The curves which are very highly peaked and have the value of $\beta_2 > 3$ are called
(a) Mesokurtic (b) Leptokurtic
(c) Platykurtic (d) All of these

81. When the distribution of the data is continuous, which distribution is applicable ?
(a) Normal distribution (b) Binomial
(c) Poisson (d) None of these

82. Normal distribution is due to the work of
(a) James Bernoulli (b) SD Poisson
(c) Laplace and Gauss (d) Demoivre

83. Binomial distribution is a very useful distribution for dealing with .... variates.
(a) Continuous (b) Discrete
(c) Both (d) None of these

84. The mean of the binomial distribution is
(a) np (b) npq
(c) $\sqrt{npq}$ (d) None of these

85. The first and third moments about the mean of normal distribution is
(a) 0 (b) σ^2
(c) $3\sigma^2$ (d) None of these

86. β_1 and β_2 constants of the normal distribution are ...., respectively
(a) 0,3 (b) 3,0
(c) 0,0 (d) 3,3

87. The arithmetic mean of the Poisson distribution is
(a) $3\ m^{2+m}$ (b) $\sqrt{m}$
(c) m (d) 1/m

88. Arithmetic mean and variance are always equal in
(a) Normal distribution (b) Poisson distribution
(c) Binomial (d) All of these

89. The hypothesis of no difference is known as
(a) Simple hypothesis (b) Null hypothesis
(c) Composite hypothesis (d) Alternative hypothesis

90. The probability of committing type I error is known as
(a) Test of significance (b) Level of significance
(c) Composite hypothesis (d) None of these

91. The statistical procedure for estimating whether the difference under study is significant or non-significant is known as
(a) Test of significance (b) Standard error
(c) Level of significance (d) Sampling

92. Which test is used to test the significance of the difference between two means ?
(a) t-test (b) F-test
(c) Z- test (d) x^2- test

93. Student's t-test is used when sample size is .... and population standard deviation is not known.
(a) Large (b) Small
(c) Both (d) None of these

94. For paired observation, which test is used for testing the significance of a mean difference ?
(a) Paired t-test (b) t-test
(c) F-test (d) x^2-test

95. For comparison of two means from independent samples which test is applicable ?
(a) t-test (b) x^2-test
(c) F-test (d) z- test

96. For test of goodness of fit, which statistical test is applied ?
(a) x^2 - test (b) F-test
(c) z-test (d) t-test

97. The value of x^2 ranges from............. and is always positive.
(a) –1 to +1 (b) 0 to 2
(c) o to ∞ (d) None of these

98. x^2 is calculated by the help of the formula
(a) $x^2 = \Sigma(O\text{-}E)^2/E$ (b) $x^2 = \Sigma(O\text{-}E)^2/(E)$
(c) $x^2 = \Sigma(O\text{-}E)^2/\Sigma O$ (d) $x^2 = \Sigma(E\text{-}O)^2/\Sigma(O)^2$

99. For calculation of x^2, N should be at least
(a) 20 (b) 30
(c) 40 (d) 50

100. For x^2-test no theoretical cell frequency should be small, it should be at least
(a) 2 (b) 5
(c) 8 (d) 12

101. x^2-test is applied for
(a) Testing of independence of attributes
(b) Genetic problem and detection of linkages
(c) Testing the expectation of a ratio
(d) All of these

102. The validity of correlation coefficient is tested by
(a) z-test (b) t-test
(c) x^2 – test (d) F-test

103. The range of correlation coefficient lies between
(a) –1 to +1 (b) 0 to 1
(c) 0 to ∞ (d) 0 to 100

104. The validity of regression coefficient is tested by
   (a) z-test (b) t-test
   (c) x^2-test (d) Both a and b

105. Coefficient of determination is a square of
   (a) Correlation coefficient (b) Regression coefficient
   (c) Both (d) None of these

106. The range of multiple correlation coefficient lies between
   (a) 0 to +1 (b) –1 to +1
   (c) 0 to ∞ (d) -∞ to + ∞

107. The significance of multiple correlation (R) is tested by
   (a) t-test (b) F-test
   (c) z-test (d) Paired t-test

108. The repetition of the treatment under investigation is known as
   (a) Randomization (b) Replication
   (c) Local control (d) None of these

109. The allocation of the treatments to the different experimental units in a random manner is known as
   (a) Randomization (b) Replication
   (c) Local control (d) Sampling

110. The principle of making use of greater homogeneity in groups of experimental units to reduce the experimental error is referred as
   (a) Replication (b) Randomization
   (c) Local control (d) None of these

111. For the expression of standard error of mean, which formula is applicable ?
   (a) $S/\sqrt{n}$ (b) $2S/\sqrt{n}$
   (c) $\sqrt{S}/n$ (d) $n/\sqrt{S}$

112. What is/are the basic principles of field experimentation ?
   (a) Randomization (b) Replication
   (c) Local control (d) All of these

113. The completely randomized design is appropriate when the experimental material is
   (a) Limited and homogeneous
   (b) Unlimited and heterogeneous
   (c) Limited and heterogeneous
   (d) Unlimited and homogeneous

114. Local control is not applied for the design
   (a) RBD (b) CRD
   (c) LSD (d) SPD

115. When the fertility gradient of the field is in one direction , the appropriate design is
(a) CRD (b) LSD
(c) RBD (d) SPD

116. When the fertility gradient of the field goes in two directions, the appropriate design is
(a) Augmented design (b) CRD
(c) LSD (d) SPD

117. In randomized block design, how many treatments can be adopted without any loss of efficiency ?
(a) 10 (b) 15
(c) 18 (d) 20

118. How many number of treatments can be adopted in LSD ?
(a) 5 to 8 (b) 5 to 12
(c) Both (d) Not estimated

119. When there are several factors with different levels to be experimented simultaneously with the same level of precision, the appropriate design is
(a) CRD (b) LSD
(c) RBD (d) Factorial scheme

120. When the experimental material is very less and number of genotypes are very large, the most suited design is
(a) Split plot design (b) LSD
(c) RBD (d) Augmented design

121. In an experiment, a treatment requires large area and b treatment requires smaller area (different precision), the most appropriate design is
(a) Split plot design (b) LSD
(c) RBD (d) Strip plot design

122. The minimum error degree of freedom should be at least
(a) 6 (b) 12
(c) 18 (d) 20

123. The ANOVA is a tool by which total variation may be split up into several physically assignable components was defined by
(a) Karl Pearson (b) RA Fisher
(c) Horace Secrist (d) AL Bowley

124. The values of regression coefficient lie between
(a) $-\infty$ to $+\infty$ (b) 0 to 1
(c) -1 to $+1$ (d) 0 to ∞

125. The correlation coefficient remains unaffected by the change of
(a) Origin (b) Scale
(c) Both (d) None of these

126. The regression coefficients are independent of change of
   (a) Origin (b) Scale
   (c) Both (d) None of these
127. When the value of correlation coefficient is 1 then two variables are
   (a) Uncorrelated (b) Partially correlated
   (c) Perfectly correlated (d) Highly correlated
128. The correlation coefficient is the geometric mean between two
   (a) Median (b) Standard deviation
   (c) Regression coefficient (d) Mode
129. The R rank correlation was suggested by
   (a) James Bernoulli (b) SD Poisson
   (c) C Spearman (d) Laplace and Gauss
130. Poisson distribution deals with
   (a) Discrete variables (b) Continuous
   (c) Both (d) None of these
131. The variation due to uncontrolled factors is spoken as
   (a) Standard error (b) Experimental error
   (c) Treatment effects (d) All of these
132. In the field experimentation, local control is used for
   (a) Validity of estimate of error
   (b) Validity of treatment effects
   (c) Diminution of error (d) All of these
133. The accuracy of measurement signifies the closeness with which an estimate approaches the
   (a) True value (b) Average value
   (c) Both (d) None of these
134. For the field experimentation the accurate idea of the fertility variation can be had by the help of
   (a) Experimental field (b) Unformity trial
   (c) Coordinated trials (d) All of these
135. Which test is applied for the test of significance between the two variances ?
   (a) t-test (b) z-test
   (c) x^2-test (d) F-test
136. If some of the counts are very small or zero, the most appropriate transformed variate will be
   (a) $\sqrt{x} + 1/2$ (b) $\sqrt{x}$
   (c) $\sqrt{x} +1/2$ (d) $\sqrt{x} +1$

137. Which transformation is most appropriate for percentages ?
(a) Square root (b) Are sine
(c) Logarithmic (d) All of these

138. If there are 5 treatments with 4 replications to each, the error degree of freedom for CRD will be
(a) 15 (b) 20
(c) 12 (d) 9

139. Which design provides maximum number of degrees of freedom for the estimation of error as compared with other design for the given number of experimental units ?
(a) RBD (b) CRD
(c) LSD (d) Factorial RBD

140. Which design is most appropriate for the laboratory experiments ?
(a) CRBD (b) CRD
(c) LSD (d) Factorial

141. Which design follows the number of rows = number of columns= number of treatments
(a) SPD (b) CRD
(c) LSD (d) RBD

142. If there are 6 levels of moisture regime and 6 replications each, the error degree of freedom for LSD will be
(a) 25 (b) 35
(c) 36 (d) 20

143. When we want to conduct an experiment on a long strip of land, which design should be preferred ?
(a) RBD (b) CRD
(c) SPD (d) LSD

144. Which design gives high precision to reduce the standard error to less than 1% ?
(a) SPD (b) CRD
(c) RBD (d) LSD

145. In confounding the precision on the main effects and certain interactions of lower order
(a) Increases (b) Decreases
(c) Remains same (d) None of these

146. If 4 levels of sowing dates are laid out in main plot, 5 levels of nitrogen applied in subplot and they are replicated three times, then what will be the degree of freedom for error (b) ?
(a) 32 (b) 6
(c) 20 (d) 60

147. Select the most appropriate design when all factors are not of equal importance in experimentation
   (a) Augmented (b) Strip plot design
   (c) LSD (d) Split plot design

148. During experimentation, we lost some information and we want to get idea about these values, which technique will be useful ?
   (a) Field plot technique (b) Missing plot technique
   (c) Seed plot technique (d) Uniformity trial

149. Indian Agricultural Statistical Research Institute is established at
   (a) Lucknow (b) Jaipur
   (c) New Delhi (d) Kolkata

150. 'Annawari' procedure is a
   (a) Traditional method for estimation of crop yield
   (b) Method for weather forecasting
   (c) Method for rainfall forecasting
   (d) All of these

151. In crop cutting survey on sugarcane recommended the plot size for estimation of crop yield is
   (a) 20m x 10m (b) 10m x 10m
   (c) 10m x 5m (d) 5m x 5m

152. Indian Livestock Census is conducted after every
   (a) 2 years (b) 5 years
   (c) 10 years (d) 20 years

153. In India the available acreage statistics for the total geographical area is
   (a) 51% (b) 61%
   (c) 71% (d) 100%

154. At present time how many number of forecasts are issued by Government of India on 34 important crops ?
   (a) 50 (b) 60
   (c) 70 (d) 80

155. Which measure of central tendency is appropriate for index numbers ?
   (a) Harmonic mean (b) Arithmetic mean
   (c) Mode (d) Geometric mean

156. Ogive is a method of
   (a) Diagrammatic representation
   (b) Graphical representation
   (c) Three dimensional diagram
   (d) None of these

157. The x^2 (chi-square)test for goodness of fit and independence of attributes come under
(a) Parametric tests (b) Non-parametric test
(c) Both (d) Not classified

158. The non-parametric test used two independent samples is
(a) t-test (b) z-test
(c) F-test (d) Mann-Whitney U-test

159. For the comparison of two correlated samples under non-parametric alternative to paired F-test is
(a) z-test (b) F-test
(c) Wilcoxon Signed-rank test (d) Kruskal-Wallis H-test

160. Under non-parametric the comparison of several related samples is tested by
(a) Quade test (b) Friedmann test
(c) Both (d) F-test

161. When both the variables are not normally distributed, in such case we may use
(a) Partial correlation (b) Multiple correlation
(c) Both (d) Rank correlation

162. Rank correlated is tested by
(a) Quade test (b) Friedmann test
(c) t-test (d) F-test

163. The relationship between independent and dependent variables is computed by
(a) Regression coefficient (b) Correlation coefficient
(c) Both (d) None of these

164. The optimum plot size in the experiment is/are determined by
(a) Maximum curvative method
(b) Fair field Smith's variance law
(c) Both (d) None of these

165. The critical difference (CD)is a form of t-test and its formula is given by
(a) CD= t. SE(d) (b) CD= t + SE (d)
(c) CD= t/ SE(d) (d) CD= F. SE (d)

166. If two factors are involved in an experimentation and both the factors require long plot size, the design suited for this will be
(a) Split plot (b) Strip plot
(c) CRD (d) RBD

167. If four levels of depth of ploughing are applied to vertical strip, five levels of moisture regimes are applied to horizontal

strip and replicated three times, then what will be the degree of freedom for error (c).[(r-1), (a-1),(b-1)] ?

(a) 8 (b) 6
(c) 10 (d) 24

168. When 4 levels of sowing dates are applied to vertical strip, five levels of irrigation are applied to horizontal strip, four levels of nitrogen randomly allocated within the intersection plot of vertical and horizontal strips and replicated three times, then what will be the degree of freedom for error (d).[ab (r-1), (c-1)] (Strip split plot design) ?

(a) 72 (b) 120
(c) 24 (d) 8

169. The square of the standard deviation is known as

(a) Variance (b) Standard deviation
(c) Mean deviation (d) Coefficient of variance

170. A Measure of the peakedness or convexity of a curve is known as

(a) Skewness (b) Kurtosis
(c) Ogive (d) Histogram

171. A distribution is said to be symmetrical if the values are uniformly distributed around the

(a) Mean (b) Mode
(c) Median (d) Mean deviation

ANSWERS

1. (c)	2. (a)	3. (d)	4. (a)	5. (a)
6. (c)	7. (d)	8. (b)	9. (b)	10. (b)
11. (d)	12. (b)	13. (b)	14. (c)	15. (a)
16. (d)	17. (d)	18. (c)	19. (c)	20. (b)
21. (b)	22. (d)	23. (b)	24. (c)	25. (a)
26. (a)	27. (d)	28. (c)	29. (c)	30. (b)
31. (c)	32. (d)	33. (a)	34. (b)	35. (a)
36. (a)	37. (a)	38. (b)	39. (b)	40. (b)
41. (b)	42. (a)	43. (b)	44. (b)	45. (a)
46. (d)	47. (c)	48. (c)	49. (c)	50. (a)
51. (c)	52. (c)	53. (d)	54. (a)	55. (a)
56. (b)	57. (b)	58. (a)	59. (b)	60. (c)
61. (a)	62. (a)	63. (b)	64. (d)	65. (a)
66. (c)	67. (b)	68. (a)	69. (c)	70. (a)
71. (c)	72. (b)	73. (b)	74. (d)	75. (a)
76. (d)	77. (a)	78. (a)	79. (c)	80. (b)
81. (a)	82. (d)	83. (b)	84. (a)	85. (a)

86. (a)	87. (c)	88. (b)	89. (b)	90. (b)
91. (a)	92. (c)	93. (b)	94. (a)	95. (a)
96. (a)	97. (c)	98. (a)	99. (d)	100. (b)
101. (d)	102. (b)	103. (a)	104. (d)	105. (a)
106. (a)	107. (b)	108. (b)	109. (a)	110. (c)
111. (a)	112. (d)	113. (a)	114. (b)	115. (c)
116. (c)	117. (d)	118. (c)	119. (d)	120. (d)
121. (a)	122. (b)	123. (b)	124. (a)	125. (c)
126. (a)	127. (c)	128. (c)	129. (c)	130. (a)
131. (b)	132. (c)	133. (a)	134. (b)	135. (d)
136. (a)	137. (b)	138. (a)	139. (b)	140. (b)
141. (c)	142. (d)	143. (d)	144. (d)	145. (a)
146. (a)	147. (d)	148. (b)	149. (c)	150. (a)
151. (d)	152. (b)	153. (c)	154. (c)	155. (d)
156. (b)	157. (b)	158. (d)	159. (c)	160. (c)
161. (d)	162. (c)	163. (a)	164. (c)	165. (a)
166. (b)	167. (d)	168. (b)	169. (a)	170. (b)
171. (a)				

QUESTION PAPERS (ARS EXAMINATIONS 1998—2005)

A.R.S.

Paper II

Subject : Horticulture—Vegetable Science (1998)

1. Explain the role of temperature, light and moisture in vegetable production. Give suitable examples. Also mention the ways of management of vegetable production under stress conditions of these factors.
2. Explain the breeding techniques and steps involved in development and production of the following :
   - (a) Seedless watermelon
   - (b) Hybrid onions
3. Explain the role of the following in vegetable breeding :
   - (a) Bt. Toxin genes
   - (b) Embryo culture technique
   - (c) Genetic markers
   - (d) Somaclonal variation

   Or

   Describe the scope and role of biotechnology in vegetable improvement.
4. Describe the hybrid tomato production technology suitable for plains of Northern/Southern Indian conditions under the following headings: Please also mention the agro-climatic regions for which the technology is described.
   - (a) Recommended hybrids.
   - (b) Nursery sowing transplanting – time, seed rate, spacing, seed treatment, etc.
   - (c) Manure and fertilizer application.
   - (d) Specific cultural management practices.
   - (e) Important diseases and their control measures.
5. Give the cultivation of fennel under following headings :
   - (a) Climate and soil.
   - (b) Sowing .
   - (c) Recommended manure application.
   - (d) Fertilizer and manure application.
   - (e) Cultivation problems.
   - (f) Diseases and their control measures.
6. Describe the production technology of sweet potato under following heads :
   - (a) Soil and climatic requirement.

(b) Recommended varieties.

(c) Planting materials and methods.

(d) Manure and fertilizer application.

(e) Insect-pests and their control measures.

7. Define term 'Plant regulators'. Write the various types of plant regulators and their applications in enhancing vegetable productivity and quality. Support your answer with suitable examples in various vegetable crops, specifying the recommendations.

8. Describe the practices and important considerations for successful production of vegetables in polyhouses in India. Give the recommendations of polyhouse production taking 3-4 most suitable vegetable crops.

Or

Write short notes on the following :

(a) Approaches for Integrated Nutrient Management in vegetable production.

(b) Transgenics in vegetable crop improvement.

(c) TPS technology for potato seed multiplication.

(d) Forcing techniques in vegetable production for higher income.

9. What are the various types of losses occurring in vegetables after harvesting? Describe the use of controlled atmosphere storage and radiation in reducing these losses.

or

Describe with specific examples the effect of pre-and post-harvest treatments in the improvement of shelf-life of vegetables.

10. What is Latin Square Design? Under which conditions can this design be used ? Give the advantages and disadvantages of this design. Also, give the analysis of variance table for this design.

or

Compute the value of correlation coefficient and regression coefficient for the following data and fit a regression equation of y on x. Also, obtain the value of y for x = 2.

x : 5	3	6	7	4
y : 3	2	5	6	4

Test the significance of correlation coefficient if the tabulated value is 2.776.

A.R.S.

Paper II

Subject : Horticulture – Vegetable Science (1999)

1. (a) "Vegetables are natural source of protective food". Justify the statement.
   (b) Analyse the causes of low productivity of vegetables in India.
2. What is the present status of export of vegetables from India? Suggest measures to boost export of these commodities.
3. List various applications of plastics in vegetables and spices. Discuss in detail the importance of plastics in water management and irrigation.
4. What are the major problems of vegetable seed industry in India? What steps will you suggest to put it on a sound scientific footing?
5. What importance have been brought out in the production of cauliflower in India? Briefly review the results of research on variety aspects.
6. How many groups of plant growth regulations are so far known to scientists? Describe the role of plant growth regulators in improvement to yield and quality of vegetables.
7. Explain the breeding techniques and steps involved in development and production of
   (a) Hybrid tomato
   (b) Hybrid cauliflower
8. Describe the importance and role of biotechnology for vegetable and spices improvement.
9. Write about the importance and scope of spices and condiments in India. Explain soil and water requirements of following spices :
   (a) Black pepper
   (b) Ginger
   (c) Turmeric
10. Explain :
    (a) Use of green house technology in boosting productivity of vegetables.
    (b) Embryo rescue technology in development of hybrids from distantly related parents.
    (c) Virus elimination through micropropagation.
    (d) Insect control through bioinsecticides.

A.R.S.

Paper II

Subject : Horticulture – Vegetable Science (2000)

1. Write short notes on any four of the following :
   (a) Seed plot technique.
   (b) Inducing femaleness in cucurbits.
   (c) Somaclonal variation.
   (d) Self-incompatibility in cole crops.
   (e) Whiptail in cauliflower.
2. Describe the role of micronutrients in vegetable production and also specify the symptoms of deficiency with suitable examples.
3. Justify that hybrid vegetable seed production is remunerative as well as provides more jobs to rural youth. Specify the areas most suitable for hybrid seed production of tomato, brinjal, capsicum, cabbage and carrot.
4. Write the breeding methodology for development of commercial hybrids in cabbage. Also write why we could not produce a single commercial hybrid in the country ?
5. Write the significant achievements made in vegetable breeding in India since 1975 with reference to new high yielding, biotic and abiotic resistant varieties.
6. "Randomized Block Design" is a most commonly used statistical design for conducting field experiments. Justify the statement.

   or

   Write short notes on the following :
   (a) Chi Square test
   (b) Mean
   (c) Median
   (d) Critical difference
7. Describe in brief the cultivation of Black Pepper along with export potential. Suggest suitable measures to boost export.

   or

   Describe in detail the method "Selling and Massing" suggested by Jones and Mann (1963) for improvement and maintenance of onion variety.
8. Describe the role and scope of low cost polygreen houses for tomato, capsicum, cucumber and muskmelon crops.

   or

   Describe the pre- and post-harvest factors affecting the storage life of onion, potato, tomato and muskmelon.

A.R.S.

Paper II

Subject : Horticulture – Vegetable Science (2003)

1. What are the cole crops grown in India? Which crops do you think economically best from farmer's point of view and justify your statement with logic ? Indicate the seasons of growing and states of the country also.
2. Give the cultural requirements of

   (a) Peas (b) Beans
3. (a) 'Vegetable cultivation is an industry', justify your statement.

   (b) Give history of vegetable seed production in India.

   (c) What is the difference between European and Asiatic types of seeds?
4. What should be the characters of good seed ?

   (i) Explain various methods of seed production of cole crops.

   (ii) Why crop inspection is essential? Mention stepwise.
5. What are the various kinds of vegetable classification? Explain any two classification methods in detail which fall from consumer point of view.
6. (i) Give the importance of post-harvest management.

   (ii) What are the factors affecting/responsible for heavy post-harvest losses in vegetables and how will you control them?

   (iii) Why vegetable is considered to be the part of Horti-culture?
7. (i) Write down the differences between vegetables and fruits.

   (ii) Winter vegetables and summer vegetables points of differences.

   (iii) Classification according to parts of plants used.

   (iv) Thermo-classification.

   (v) Classification based on soil acidity.

   (vi) Classification based on climatic requirements.

   Explain its importance with reference to vegetable growing in India.

8. Give the importance of potato from food point of view in India and abroad. Give its cultural requirements in detail also.
9. Differentiate between :
   (a) Heat and temperature.
   (b) Macro- and Micro-nutrients (importance).
   (c) Kitchen gardening and truck gardening.
   (d) Contact and systemic herbicide.
   (e) Antagoistic and synergestic of soil nutrition.
10. List the vegetable crops grown through nursery and how you to proceed to grow healthy nursery.

A.R.S.

Paper II

Subject : Horticulture – Vegetable Science (2005)

1. What is the present status of vegetable production in India? Describe the importance of vegetables in nutritional security and national economy. How India can benefit from GATT and WTO agreements.
2. What is organic farming? Briefly explain its scope in India in the context of IFOAM objective and explain the components of organic farming.

   Or

   What do you mean by precision farming? Explain the production technology of capsicum and cucumber under precision farming.
3. (a) Outline the classification of *Lycopersicon* species with chief distinguishing characteristics for each genera.

   (b) List important genera in *Cucurbitaceace* family along with distinguishing key characters of each genera.

   Or

   Describe the recent techniques of biotechnology used for improvement of different vegetable crops.
4. Describe in brief the origin, soil, climate and nutritional requirements of potato and sweet potato along with two important varieties of potato in each early, medium and late maturity groups with their special characters.

5. Outline breeding objectives, scientific name, family, chromosome number, ploidy level and appropriate breeding methods for the following vegetables :

(a) Saffron (b) Ginger

(c) Coriander (d) Black pepper

(e) Cardamom

6. Explain different types of male sterility. Give suitable schemes for their maintenance and use in hybrid seed production in each type.

Or

What is the present status of hybrid seed production in India? What are the various genetic mechanisms being utilized to develop low cost F_1 seed in some vegetables?

7. List important uses of growth regulators in vegetables, citing suitable examples and explain in detail their role in sex expression of cucurbits.

8. What is plasticulture? Describe how plastics have been commercially exploited in high tech vegetable production in India.

Or

Write short notes on the following :

(a) Vegetable floating gardens.

(b) Physiological disorders of cauliflower.

(c) Missing plot technique.

(d) Terminator gene.

(e) Purning and staking in tomato.

9. What do you mean by maturity indices? Give maturity indices for tomato, muskmelon, cauliflower, peas and black pepper including the advances of assessment of maturity.

10. What are the basic principles of designing experiments? Describe briefly the main features, merits and demerits of split plot design. Considering three main plot and five subplot treatments give a layout plan and skeleton of analysis of variance.

Or

Define briefly the following terms and also comment on their usefulness in experimentation and statistical interpretations.

(a) Analysis of variance. (b) Mean.

(c) Replication. (d) Standard error.

LIST OF IMPROVED VARIETIES OF VEGETABLES

Name of Vegetable	Varieties
A. Varieties	
1. Tomato	
Determinate	HS-101, HS-102, HS-110, Sweet-72, Punjab Chhuhara, Roma, KS-1, Punjab Keshri, Sel-7, Sel-32, Sel-152, Sel-12, Pusa Early Dwarf, Pusa Gaurav, CO-1, CO-3, Labonita, Sel-1-6-1-4, Hisar Lalima, Pusa Red Plum, Pusa Sel-4, Pusa Sel-8, Pusa120, Balkan, Pusa Sadabahar.
Indetermine	Best of All Pant Bahar, Pusa Ruby, Sel-120, Sioux, T-1, Arka Vikash, Arka Saurabh, DT-10, BT-12, Pant T-3.
2. Brinjal	
Long	Pusa Purple Long, PBR-128-5, PBR-61, Sel-1, Punjab Sadabahar, Neelam, PH-4, Pusa Kranti, Pant Samrat, Azad Kranti, ARU 2C, BB-26, Sel-4, Punjab Barsati, KT-4, NDB-25, ARUK, Arka Sheel, Punjab Chamkeela, Arka Navneet, PBR-9-1, PBR-91-2, Punjab Bahar.
Round	Jamuni Gol Baigan, Pant Rituraj, H-8, KS-224, ABI, K-202-9, DBR-31, DBR-8, T-3, Pusa Purple Round.
Small Round	PLR-1, Aruna, DBSR-44.
Brinjal Green	Arka Kushmakar, Arka Shirish, Ramnagar Giant, Rajendra Baigan
3. Chilli	Andhra Jyoti (G-5), Bhagyalaxmi (G-4), K-2, K-1, Sel-1, K-1, Sel-1, J-218, Muslabadi, LCA-206 B, JCA 283, Phuley C-5, Bhaskar, CO-1, CO-2, MDU-1, NP-46-A, Punjab Lal, Pusa Jawahar, Pant C-1, Pant C-2, JCA-154, Kalyanpur Selection, Pusa Sadabahar, X-235, Kalyanpur Red, Chamatkar, Kalyanpur Red, ApamaG-3, Azad Mirch-1.
4. Capsicum (Bell Pepper)	Califomia Wonder, Yolo Wonder, Kingod North, Early Giant, Bull Nose, Chinese Giant, World Beater, Arka Mohini, Arka Gaurav, Arka Basant.

5. Okra	Pusa Sawani, Punjab No-13, Red Bhindi, CO-1, Perkins Long Green, Ghana Red, Gujarat Bhindi-1, Selection-2, EMS-8, Azad B-1, Azad B-2.
6. Watermelon	Arka Manik, Asahi Yamato, Pusa Bedana, Durgapur Kesar, Sugar baby, Durgapur Meetha, Improved Shipper, Special No.-1.
7. Muskmelon	Pusa Sharbati, Pusa Madhuras, Hara Madhu, Punjab Sunehari, Sel-2, Arka Jeet, Arka Rajhans, Lucknow Safeda, Durgapur Madhu, RM-43, IIHR-352, IIHR-190, Edisto, Honewdew.
8. Long melon	Arka Sheetal, Punjab Long Melon
9. Cucumber	Japanese Long Green, Straight Eight, Pionsette, Poona Khira, Balam Khira, Swarn Poorna, DVRM-1, IIHR-77-1, Kalyanpur Green, Pusa Uday.
10. Pumpkin	Arka Suryamukhi, Pusa Vishwas, CM-14.
11. Summer Squash	Hisar Sel-1, Patty Pan, Early Yellow Profile, Australian Green, Punjab Chhappan Kaddu-1.
12. Winter Squash	Arka Chandan, CO-1, CO-2.
13. Bottle gourd	Pusa Summer Prolific Long, Pusa Summer Prolific Round, Pusa Naveen, Arka Bahar, CO-1, Punjab Round, Punjab Long, Punjab Komal, NDBG-1, KBG-13, Azad Nutan.
14. Bitter gourd	Coimbatore Long, Arka Harit, VK-1, Priya, PMC-84, MDU-1, CO-1, Konkan Tara, Pusa Do Maushmi, Kalyanpur Sona, Kanpur Barahmasi, Pusa Vishesh, Punjab-14, NDB-1, Phule BG-6.
15. Ridge gourd	Pusa Nasdar, CO-1, Arka Sumeet, Konkan Harita, PKM-1, Punjab Sadabahar, Satputia
16. Sponge gourd	Pusa Chikni, Kalyanpur Chikni, Pusa Sneha.
17. Tinda	Arka Tinda, Punjab Tinda, S-48
18. Ash gourd	CO-1, CO-2, Pusa Shakti.
19. Snake gourd	CO-1, CO-4, TA-19

20. Pointed gourd	Swarna Alaukik, Swarna Rekha, FP-1, FP-3, FP-4.
21. Pea	Ageta-6, Arkel, Early Badger, Little Marvel, Early December, Bonneville, T-19, Lincoln, Dewiche Commado, NP-25, Perfection New Line, Thomas Laxton, Alderman, GC-141, GC-195, Kawari, Sylvia, Harbhajan, GC-447, Pant Uphar, P-88, VL-3, JP-4, Azad P_1, Azad P_2, Azad P_3 Azad P_4.
22. French Bean	Bountiful-Plantifur, Contender, Premier, King Green, Blue Lake, Kentucky Wonder, Arka Komal, VL Bouni-1, Pant Anupama, Parvati, Bonai, Phule Surekha, Azad Rajma-1.
23. Cowpea	Pusa Phalguni, Pusa Barsati, Pusa Dofasli, C-152, S-1552, S-203, Arka Garima, Sel-2-1, Red Seeded selection, T_{59}, Pusa Komal.
24. Dolichus bean	Pusa Early Prolific, JDL-37, Kalyanpur T-1, Arka Jay, Arka Vijay.
25. Cluster bean	Pusa Navbahar, Sharabdhara, P-28-1-1, CP-78, CO-160-1, Durgapur Safed.
26. Sem	CO-1, CO-2, 125-36 Wal Konkan-1, Hebbal Avare-3, Rajani, Pusa A4.
27. Cauliflower	Punjab Kunwari, Pusa Deepali, Improved Japanese, Pusa Synthetic, Snow ball-16, K-1, Early Kunwari, Pant Gobhi-3, Pusa Early Synthetic, Pant Gobi-2, PG-26, PG-26, PG-35, Hisar-1, IIHR-101, IIHR-105, Pant Subhra, Pusa Subhra, Pusa Himjyoti, Pusa Katki, Pusa Snowball-1, Pusa Snowball-2, Pusa Snowball K-1.
28. Cabbage	Golden Acre, Pusa Mukta, Pride of India, Copenhangen Market, Pusa Drumhead, Late Large, Drum Head, Red Cabbage, Golden Eve.
29. Knol-khol	White Vienna, Purple Vienna, King of North, Early White Vienna, Large Green.
30. Radish	Pusa Chetki, Pusa Desi, Pusa Reshmi, Japanese White, Punjab Safed, Kalyanpur No.-1, CO-1, Arka Nishant, Chinese Pink, HR-1, Red Tail Radish, Palam Hriday.

31. European type radish	Pusa Himani, White Icicle, Rapid Red White tipped, Scarlet Globe, Scarlet Long.
32. Turnip (Asiatic type)	Pusa Kanchan, Pusa Swati, Punjab Safed-4, Early Milan Red Top.
33. European (Temperate type)	Purple Top, White Globe, Goldenball, Snowball, Pusa Chandrima, Pusa Swarnima.
34. Carrot (Asiatic type)	Pusa Keshar, Pusa Meghali, Sel-233, No-29.
35. European type	Nantes Half Long, Early Nantes, Chantenay, Imperator Zeno, Pusa Yamdagni.
36. Onion (Red)	Agri Found Dark Red, Agri Found Light Red, Pusa Red, Punjab Selection, Pusa Ratnar, N-2-4-1, Arka Niketan, Arka Kalyan, Pusa Madhvi, Punjab Red Round, Kalyanpur Red Round, Hisar-2, Arka Pragati, N-53, Udaipur-101, Udaipur-103, VL-3, Arka Bindu-780, Agrifound Red Rose, Kalyanpur Red Round.
37. Onion (Yellow)	Early Grano, Spanish Brown.
38. Onion (White)	Pusa White Flat, Pusa White Round, Udaipur-102, Punjab-48, N-257-9-1.
39. Garlic	Agrifound White-G-41, Yamuna Safed (G-1), G-50, G-282, Agrifound Parvati (G-313), IC-49373, IC-49382, T-86/Sel-1, EC-198250.
40. Amaranth	Pusa Early Bunching, Bari Chaulai, CO-1, CO-2, CO-3, CO-4, Pusa Kirti, Pusa Kiran.
41. Spinach	All Green, Palak Green, Pusa Joyti, Pusa Harit, Jabner Green, HS-23, Palak No 51-16.
B. Hybrids	
42. Brinjal (Long)	Pusa Hybrid-5, NDBH-6, NDBH-7, HOE-404, Pant Hybrid-1, ARBH-201, HOE-414.
43. Brinjal (Round)	Pusa Hybrid-6, Pusa Hybrid-9, ARBH-216, NDBH-1, Neembaker, Pant Hybrid-2, BH-1, BH-2, HOE-44, CHRBH-1, CHRBH-2.
44. Brinjal (Small Round)	MHB-10, MHB-39, ABH-1, ABH-2, Hybrid-2, Sumex-9, Sumex-19, HOE-4.

45. Tomato (Determinate)	Pusa hybrid-2, NA-501, DTH-4, NDTH-1, Phule Hybrid-1, BSS-39, ARTH-3, NA-701, ARTH-15, Hybrid No.37, Swarna-12, NDTH-6, HOE-606, NA-701, Maitri, Rishi, HOE-616, Pusa Divya.
46. Tomato (indeterminate)	ARTH-4, MTH-6, NA-601, Arka Vardhan, KT-4, FMH-1, BSS-20, BSS-40, BSS-90, HOE-909, Larica Ratna, DTH-6, Sonali, ARTH-16, FM-2, NDTH-2, NDTH-4, TC-161.
47. Chilli	Agni, HOE-808, HOE-888, HOE-818, CH-104, BSS-138, BSS-141, CH-1 (MSIs x LSS), ARCH-228.
48. Capsicum	Bharat, Early Bonty, Indira, HOE-801, Hira.
49. Cabbage	Pusa Synthetic, Sri GaneshGol, BSS-32, Quisto, Nath-401, NATH-501, Uttam.
50. Cauliflower	Pusa Hybrid-2.
51. Watermelon	Arka Jyoti, Nath-102, MHW-6, Nath-202, NHW-11.
52. Muskmelon	Punjab Hybrid-1, MH-10, Pusa Rasraj (M-3), DMH-4, MHY-3, MHY-5.
53. Pumpkin	Pusa Hybrid-1.
54. Cucumber	Pusa Sanyog, Solan Hybrid-1, PCUC-1-1, AAUC-1, AAUC-2.
55. Squash	Pusa Alankar.
56. Bottle gourd	Pusa Manjari, Pusa Meghdoot, NDBH-7, NDHGH-4, Pusa Hybrid-2, PBOG-2, PBOG-1.
57. Carrot	Hybrid-1.

LIST OF VEGETABLE RESEARCH CENTRES IN INDIA

Vegetable Crops	Centres
1. Amaranth	NBPGR (New Delhi) and Coimbatore
2. Brinjal	Hessarghatta, Bhubaneswar, Kalyani and Sabour
3. Capsicum	Katrain and Solan
4. Cabbage	Katrain and Kalpa (H.P.)
5. Carrot	Hissar (Tropical) and Srinagar (Temperate)
6. Cauliflower (Early)	Sabour and Faizabad
Cauliflower (Mid-Season)	Pantnagar, Sabour and IARI
Cauliflower (Late)	Katrain and Solan
7. Chillies	Coimbatore and Lam
8. Cucumber	Kalyani and Rahuri
9. Dolichos bean	NBPGR (New Delhi), Jabalpur and Kalyanpur
10. French bean	Hessarghatta, Rahuri and Solan
11. Garlic	NBPGR (New Delhi), AADF (Nasik) and Solan
12. Muskmelon	Durgapura, Faizabad and Ludhiana
13. Okra	NBPGR (New Delhi), Hessarghatta and Bhubaneswar
14. Onion	NBPGR (New Delhi), Rahuri and AADF (Nasik)
15. Pointed gourd	Sabour, Kalyani and Faizabad
16. Peas	IARI (New Delhi), Jabalpur, Kalyanpur and Ludhiana
17. Pumkin	Hessarghatta and Faizabad
18. Tomato	NBPGR (New Delhi), Hessarghatta, Ludhiana and Kalyanpur
19. Watermelon	Durgapura and Faizabad

STATE-WISE VEGETABLE RESEARCH STATIONS IN INDIA

UTTAR PRADESH

Vegetable Farm Kalyanpur, Kanpur.
Potato Research Farm, Farrukhabad.
Horticultural Research Station, Saharanpur.

UTTARANCHAL

Potato Research Station, Kausambi, Almora.

ANDHRA PRADESH

Agriculture Research Station, Lam., Guntoor.
Demonstration Agricultural Station, Araku Valley, Vishakhapattam.
Government Main Agricultural Station, Himayatsagar, Hyderabad.
Government Model Orchard-cum-Nursery, Giddalpur, Kurnool.
Government Fruit Farm, Ramapachodavaram, Rajamundry.
Government Fruit and Vegetable Farm, Sangareddy, Medak.

ASSAM

Agriculture Research Laboratory, Jorhat, Sibsagar.
Ginger Research Station, Nayabunglow, Khasi Hills.
Horticulture-cum-Composite Research Station, Dergaon Sibsagar.

BENGAL

Agriculture Research Institute, Tollyganj.
Horticultural Research Station, Krishangar, Nadit.
Potato Research Station, Bhanjang, Darjeeling.

BIHAR

Agriculture Research Institute, Darbhanga.
Agriculture Research Institute, Patna.
Potato Breeding and Research Station, Patna.

MAHARASHTRA

Agriculture Research Station, Karad, North Satara.
Agriculture Research Station, Niphad, Nasik.
Agriculture College Vegetable Farm, Poona.

HIMACHAL PRADESH

Potato Development Station, Shilaroo, Mahasu.
Potato Development Station, Ahla, Chamba.
Potato Research Station Kufri, Shimla.
Vegetable Sub-Station, Katrain, Kullu Valley.
Agriculture Farm, Parala, Mahasu.

JAMMU AND KASHMIR

Provincial Experimental Farm, Shalimar, Shrinagar.
Provincial Experimental Farm, Golsamandar, Shrinagar.

KARNATAKA

Government Agricultural Farm, Ponnampet.
Regional Farm, Yelavare, Harnahatti, Ariskere Taluk.
Tungabhadra Agricultural Research Station, Siruguppa, Bellary.

KERALA

Agriculture Research Station, Ambalavayal, Cannanore.
Agriculture Research Station, Taliparamba, Cannanore.
Agriculture Research Station, Pattambi, Palghat.
Pepper Research Station, Panniyur.

MADHYA PRADESH

Government Experiment Farm, Chhindwara.
Government Research Farm, Kuthalia, Rewa.
Government Seed and Demonstration Farm, Waraseoni, Balaghat.
Government Seed cum Demonstration Farm, Kaliprada, Kararia, Bhopal.
Horticultural Research Station, Pachmadi, Hosangabad.

TAMIL NADU

Agriculture Research Station, Aduturai, Thanjore.
College of Orchard Farm, Coimbatore.
Pomological and Vegetable Station, Nilgiris.

MANIPUR

Government Agricultural Farm, Imphal.

ORISSA

Agriculture Research Station, Sambhalpur.
Khurda Farm, Khurda, Puri.
Khurda Farm, Kudra, Cuttack.
State Research Station, Bhubaneswar, Puri.
Sukinda Farm, Sukinda, Cuttack.
Turmeric Research Station, Gudayagiri, Boudh Phubani.

PUNJAB

Tuber Crops Research Station, Jalandhar.
Vegetable Research Station, Jalandhar.
Bhupendra Agriculture Farm, Raeeni, Patiala.

RAJASTHAN

Plant Breeding and Vegetable Research Farm, Durgapura, Jaipur.

ABBREVIATIONS

AEZ	Agri-Export Zones
AIFPA	All India Food Preserver's Association
APEDA	Agricultural and Processed Foods Product Export Development Authority
ARIC	Agricultural Research Information Centre
ATIC	Agricultural Technology Information Centre
AVRDC	Asian Vegetable Research and Development Centre
CAU	Central Agricultured University
CFB	Corrugated Fibre Board
CFTRI	Central Food Technological Research Institute
CHES	Central Horticulture Experimental Station
CIAH	Central Institute for Arid Horticulture
CIS	Crop Insurance Scheme
CIPHET	Central Institute for Post Harvest Engineering and Technology
CISH	Central Institute for Sub-Tropical Horticulture
CITH	Central Institute for Temperature Horticulture
CPRI	Central Potato Research Institute
CPRS	Central Potato Research Station
CRL	Central Research Laboratory
CSC	Central Seed Committee
CTCRI	Central Tuber Crops Research Institute
CVRC	Central Variety Release Committee
FPTC	Food Processing and Training Centre
GMS	Genetically Modified Seed
GMF	Genetically Modified Food
GW	Global Warming
HETC	Horticulture Experiment and Training Centre
HOPCOMS	Horticulture Producers Co-operative Marketing Society
HPMC	Horticultural Produce Marketing Co-operation
HYVP	High Yielding Varieties Programme
IAHSCO	Indo-American Hybrid Seeds Company
IFM	Integrated Fertility Management
IIHR	Indian Institute of Horticulture Research
IIPR	Indian Institute of Pulse Research
IIVR	Indian Institute of Vegetable Research
IPC	International Potato Centre

IPRs	Intellectual Property Rights
IPGRI	International Plant Genetic Resources Institute
IPM	Integrated Pulse Management
ISHS	International Society for Horticultural Science
IVLP	Institute Village Linkage Programme
IWM	Integrated Weed Management
KCC	Kisan Call Centres
KRIBHCO	Krishak Bharti Co-operative Ltd.
KGK	Krishi Gyan Kendra
KVK	Krishi Vigyan Kendra
LAN	Local Area Network
MAHYCO	Maharashtra Hybrid Seeds Company Ltd.
NAS	National Agriculture Scheme
NATP	National Agricultural Technology Project
NAIP	National Agriculture Inovative Project
NBPGR	National Bureau of Plant Genetic Resources
NCF	National Commission for Farmers
NFPP	National Food Processing Policy
NHB	National Horticulture Board
NDP	National Demonstration Project
NOVDB	National Oilseed and Vegetable Development Board
NPA	National Policy on Agriculture
NSC	National Seeds Corporation
NVT	National Varietal Trials
PBCS	Potato Breeding and Certification Station
PBRs	Plant Breeders Rights
PF	Precision Farming
PURA	Providing Urban Amenities in Rural Areas
SAUs	State Agricultural Universities
SSDC	State Seed Development Corporation
SVRC	State Variety Release Committee
TRIPs	Trade Related Intellectual Property Rights
TT	Terminator Technology
TTC	Trainers Training Centre
UVT	Uniform Varietal Trials
VPKAS	Vivekananda Arvatiya Krishi Anusandhan Sansthan
WWW	World Wide Web
WTO	World Trade Organization
GATT	General Agreement on Tariff and Trade